VOICES
FROM THREE
MILE
ISLAND

This book is dedicated
to every mother, every father, every child;
to all the healers and life-builders
on this planet.
Let us heed the warning and
act now,
before it is too late.

VOICES

FROM THREE MILE ISLAND

The People Speak Out

BY ROBERT LEPPZER

 THE CROSSING PRESS / Trumansburg, New York 14886

Acknowledgments:

I wish to thank the following people for their help and contributions to this work. First and foremost, my appreciation goes to my true friend and ally, Al Giordano, the singing fire-brand/"No Nukes Troubadour" of the northeast, whose original idea and energy led four New Englanders (the two of us with Eleanor Swaim and Sue Stitely of Vermont to venture down to Pennsylvania last January for a people-to-people fact finding tour. Many thanks to Kay Pickering and Chris Sayer of Three Mile Island Alert, whose invaluable aid led us to many of our contacts; Eleanor Green of Lancaster, Penn. for putting us up in her house during our stay; WMUA, the staff of WFCR (Amherst), National Public Radio, Pacifica Program Service, Jim Koplin and the UMass/Amherst chapter of SCANN (Students Against Nukes Nationwide) for their aid in insuring a wide distribution of the original radio documentary; Vince Darago and the Public Interest Video Network, for their support in helping me to record interviews in March; Donn Young and Bill Kiesling for their fine photographic work in this book; special thanks goes to John Gill of the Crossing Press, whose vision, encouragement and trust in me has made this book a reality; and to many close friends whose support and encouragement has helped me greatly through this long, arduous endeavor.

I must give a final word of appreciation and admiration to the people of Three Mile Island interviewed in this book for their warm hearts and courage to speak out and tell their stories so that others may understand and learn from their ordeal.

Copyright © 1980 by Robert Leppzer

Cover and Book Design: Mary A. Scott

Photo Credits: Front Cover, pages 19, 26, 31, 40, 49, 69 by *Valley Advocate*/Donn Young. Pages 3, 8, 13, 35, 43, 53, 59, 64, 75 by Bill Keisling. Page 82 by Barry Pearl Studio.

Printed in the U.S.A.

Library of Congress Cataloging in Publication Data
Main entry under title:

Voices from Three Mile Island.

1. Three Mile Island Nuclear Power Plant, Pa.
2. Atomic power-plants--Accidents--Social aspects--
Pennsylvania. 3. Atomic power-plants--Environmental
aspects--Pennsylvania. I. Leppzer, Robert.
TK1345.H37V64   363.1'79   80-20933
ISBN 0-89594-041-8
ISBN 0-89594-042-6 (pbk.)

Introduction

On March 28, 1979, the worst accident in the history of the commercial nuclear power program in the United States occurred in Unit 2 of the Three Mile Island nuclear plant, just twelve miles outside of Harrisburg, Pennsylvania. The most feared reactor accident, a core meltdown, which the nuclear industry has continually assured the public could never happen, almost *did* happen. A report by the Nuclear Regulatory Commission concluded that the Three Mile Island plant came within thirty to sixty minutes of a full meltdown. In fact, the core of the reactor did undergo partial melting.

What are the consequences of a nuclear meltdown? In the mid 1960's, the Atomic Energy Commission reported that, should a meltdown ever happen, it would kill thousands of people outright from a lethal dose of high level radiation; expose hundreds of thousands of people to low level radiation which would bring dramatic increases in leukemia, cancer and birth defects within five to thirty years; and contaminate a land area the size of Pennsylvania for over one hundred years.

How much radiation were the people living near the plant exposed to? No one knows for sure. The operators of the plant, Metropolitan Edison Company, and the Nuclear Regulatory Commission concur that the emissions at the time of the accident were higher than normal but still posed no danger to nearby residents.

However, one of the many scientists who are critical of the nuclear industry, Dr. Ernest Steinglass, Professor of Radiation Physics at the University of Pittsburgh, states that the actual doses of radiation to human beings' internal organs were "thirty to one hundred thirty times greater than reported by the NRC to the public."

The accident at Three Mile Island is not yet over. More than six hundred thousand gallons of highly radioactive water still must be disposed of. In March, 1980, statements by NRC Commissioner, Joseph Hendrie, revealed new concern that the pipes, valves and gauges inside the plant may be deteriorating. One member of a special NRC taskforce was quoted as saying, "There's actually no way to tell whether pipes now under water in the containment building will hold up for years, or go very soon."

The book you are about to read is not about statistics and figures —it is a very personal one. It is the story of America's near catastrophic nuclear accident as told by the people who lived through it— and who must live with its aftermath. It is a montage of voices expressing the human side of this traumatic experience—fear, anger, frustration and betrayal. Presented here are the accounts of people who live within a few miles of the Three Mile Island nuclear plant, told in their own words, without editorial comment. The material was drawn from over fifteen hours of interviews conducted on location in January and March, 1980, with additional material provided by phone interviews through July, 1980.

Much of this material was broadcast as a two hour radio documentary special which I produced, EARLY WARNINGS: VOICES FROM THREE MILE ISLAND. It was aired by over forty-five stations around the country, including thirty National Public Radio affiliates, to commemorate the first anniversary of the Three Mile Island accident in March, 1980.

The People

FRAN CAIN lives directly across the street from Three Mile Island. Her house on Meadow Lane in Middletown is approximately one thousand feet from the containment building of Unit 2 where the damaged reactor core lies. Thirteen years ago, at the age of forty, Fran embarked on her career as a professional dog groomer and breeder of high quality toy poodles. She operates her business out of her home.

CHARLES CONLEY is still farming at seventy-four years old. He lives less than two miles from TMI in Etters. He farms a four-year rotation of corn, oats, wheat and alfalfa. On the side, he also grows some potatoes and pumpkins.

ELIAS CONLEY has been farming all his life in Etters. At age seventy-one, he maintains a one hundred twenty acre farm, growing corn, oats, wheat and hay. He and his wife, Gladys, live about three miles from the nuclear plant.

VICKIE DISANTO was born in Middletown and has lived there for most of her life. She presently lives less than a mile from the TMI reactors. She's married and has two children—Albert (T-Bird), age two, and Angela, born in excellent health on May 6, 1980.

VANCE FISHER farms one hundred acres of farmland, three miles from TMI in Etters. Corn, barley, wheat and hay are among his crops. Born and raised on this farm, he has lived most of his sixty years in this farming valley on the west shore of the Susquehanna river.

DR. MICHAEL GLUCK works out of three hospitals in the Harrisburg area as a general practioner. He did his residency in internal medicine at the Harrisburg Hospital, where he was Chief Resident in 1977. He's married and has two children. At the time of the accident, he and his family lived in Newberry Township, about two miles from the TMI plant.

DR. JUDITH JOHNSRUD is co-director of the Environmental Coalition on Nuclear Power, a state-wide organization of citizen groups in Pennsylvania that has been active fighting nuclear power since 1968. Since 1973, she has been one of the legal intervenors on behalf of the Coalition in the licensing proceedings before the Nuclear Regulatory Commission on Three Mile Island, Units 1 and 2. In addition, she is active with the Pennsylvania Governor's Energy Council Advisory Committee and is the Vice-Chairperson of the National Solar Lobby. Dr. Johnsrud is a geographer by profession.

JANE LEE is the assistant manager of the Jeramiah K. Fisher dairy farm in Etters, three-and-a-half miles from TMI. This farm has been in the same family, generation after generation, since the early 1700's. Among her farm duties, Jane drives a tractor, helps to bail hay and milks the cows. She has lived on the farm for over nine years as a friend of the family, with Jerry and Joan Fisher and their eight year old son, David.

CATHIE MUSSER lives in Harrisburg. However, at the time of the accident she lived in Londonderry Township, just five miles from the plant. She is married and the mother of a two year old girl and is active in Three Mile Island Alert, a Harrisburg anti-nuclear group.

BOB REID is the Mayor of Middletown, Pennsylvania. He has lived in this town for all of his forty-seven years. "I've always been interested in politics," he says. Reid served ten years on the city council and has been mayor for three years. He has been a teacher in the Middletown school system for twenty-one years. For the past nine years, he has taught a high school course in government.

PAT STREET lives with her husband, Richard, and her two children—Jenny, age seven, and Michael, age nine—less than a mile from Three Mile Island in Middletown. She has lived most of her life in the Harrisburg area, living for the past eight years in Middletown. At the time of the accident, she was working as a bookkeeper in a nearby law firm.

DR. ROBERT WEBER has been a practicing veterinarian since 1947. "I accept any animal that needs treatment," he says. The majority of Dr. Weber's patients are large farm animals such as dairy cows, steers, horses, hogs and sheep. In addition, he treats quite a few smaller farm animals including ducks, geese, turkeys, and pets such as dogs, cats, skunks, raccoons, possums, groundhogs—whatever people choose to have as pets. "Sometimes they even bring in snakes," he adds. His coverage area extends to four counties in a twenty mile radius from his office in Mechanicsburg, fifteen miles from Three Mile Island.

BILL WHITTOCK lives on the Susquehanna river one mile across from the Three Mile Island nuclear plants in Goldsboro, Pennsylvania. He has lived most of his seventy-four years in the nearby area, residing in his present house for ten years. As a civil engineer by profession, he served as a Lt. Commander overseas in the Navy Construction Battalion in World War II. After the war, he established his own consulting engineering firm and was owner of this until his retirement several years ago.

The Accident

BILL WHITTOCK:
The accident occurred around four o' clock in the morning on Wednesday, March 28. I was sleeping and became aware of this explosive release of steam over at the plant. There was a roar, a *terrific* roar. It woke me up and I jumped out of bed. I looked across the river and saw the column of steam that was escaping. It roared for about five minutes. It stopped and then it started to roar again.

After the second time, it eased off until there was just a relatively small column of steam which just kept hissing pretty near all morning. I went back to sleep because the plant had erupted about ten times before and I thought it was just an ordinary disturbance like we've been going through for the last several years.

About seven o'clock when I got up I heard on the radio that there had been a radioactive release. Then I went uptown to get the mail. I could sense a metallic taste in the air when I got outside. I asked up at the marina if the people up there had sensed this taste. Two of them did.

About nine-thirty a helicopter came in and landed up here in the field. There was a TV crew that came down. That was when I began to realize that it was pretty serious.

MAYOR BOB REID:
I was called by my civil defense director who told me it was an accident. I asked him, "Well, what kind of accident is it, Butch?" And he said, "Well, all we've gotten so far is an *on-site emergency*."

We had no concrete information that we could go on. We had a television set in our communication center but each channel gave us different information. We had a radio but each station gave us different information. We didn't know what to do. So I called the home office of Metropolitan Edison Company. They told me that "Yes, there was an accident" and tried to explain to me just exactly what took place. I said, "Well look, I don't want to hear the technical aspects of this darn thing because I don't understand it. What about radiation?" I knew that. He said, "Oh, no radiation was released. You don't have to worry about that. No radiation was released and no one was injured." I said, "Great!"

So I told Butch, "Well look, I'm going back to school. If anything else happens let me know." I turned my radio on in the car and the first thing the announcer said was that radiation *was* released. I'd just talked to an official from the plant at eleven o'clock. At four o'clock in the afternoon the same man called me and said, "Bob, I'd like to update our conversation." I said, "You're going to tell me now that radiation was released." He said, "Yes." I said, "Well, I guess we're in for a lot of malarkey from you people." And lo and behold, boy, there it was. At four o' clock on the 28th. I didn't trust Met. Ed.

PAT STREET:
I was at work and I heard a little blurb over the radio, "There's been a release at Three Mile Island." No big deal. We've had these before. This is routine. I didn't think anything about it. My family came and we were joking about it. "Are you sure it's all right to be here? Ha, ha, ha." We got a phone call from my in-laws in Maryland. *"What are you still doing there?* Aren't you coming here?" "What for? We've had releases before. No, we're staying here. There's nothing to worry about." I really wasn't worried.

VICKIE DISANTO:
The morning that it happened I had no idea until my mother called me from work to tell me that there was an accident. I kept waiting for someone to tell me get out or *do* something. I listened to all the news reports but I didn't hear anything. Finally about six o'clock that night, we did hear a report on the news. So we called all of our

Middletown Mayor Robert Reid

3

relatives in California and told them that everything was okay so that they wouldn't be upset.

JANE LEE:
The first reports were very superficial and presented in a way that wouldn't create panic. But I realized with all the experience that I had with those plants down there that something was *very, very* wrong. I got home about noon time and I called Judy Johnsrud up at State College who had been active fighting the plant for years. I said, "All right, tell me what's going on at TMI." She said, "Jane, you're not still there!" I said, "What do you mean?" And she said, "You better get out of there. That plant's on its way to a meltdown." I said, "You're kidding me." And she said, "No, I'm not kidding. You better get the hell out of there."

Well there I was. I had to tell Joan and David and Jerry. We had 85 head of cattle out there, plus goats, dogs, cats and ducks. What in the name of God do you do? I was upset, but I wasn't in a state of panic because I figured it had happened at four o'clock and I was still here. But I didn't know how much we were exposed to.

DR. MICHAEL GLUCK:
My wife heard it on the radio that day. She was about six months pregnant at the time. We also had a one year old at home. There were reports on the first day that no radiation was getting out at all. At six o'clock that night the national news and the local news said two different things. The national news said that they didn't know how much radiation was getting out. The local news said all the radiation was contained and there was no danger at all. At that point we felt that our house was unsafe and we left for Hershey. We slept at a friend's house and the next morning we went 100 miles away. We didn't wait at all.

BILL WHITTOCK:
On Thursday, I decided to leave. Fortunately, my wife and daughter were away. I called them on the phone and told them. I took my animals and went up through town. I saw many people in town hastily getting ready to vacate. About sixty percent of the town left then. Some people went as far as Florida, some went out to

Pittsburgh. I didn't go more than fifteen miles away because I had an apartment there. I stayed there for about ten days.

I was scared. If I hadn't been scared I wouldn't have left. I think everybody felt the same way. Some people left more to protect their children than themselves. I was afraid that we might get a meltdown and a lethal release that might be fatal.

MAYOR BOB REID:
There was no panic but you could see people were concerned. They had a sense of humor but it was the type where you see people laughing and crying at the same time. They were trying to be brave. And I felt that they really *were* brave.

VANCE FISHER:
I stayed here when the TMI problem arose. The rest of the family went to visit some friends in York so that at least they would have a headstart if it got real bad. But what with the animals and all the equipment, I thought I should stay around and keep an eye on them. I really wasn't that much concerned with radiation.

I was a little confused about the hydrogen bubble they kept speaking about. They were a little scarce with the information they gave us. I thought, well, it's quite possible the hydrogen could explode and things would get rough. Of course, I had hopes that before it got too bad I'd be able to hold my breath and get out of the area.

PAT STREET:
We heard over the news, "The Governor is considering evacuation." You know, the first thing that goes through your head is—my family. I tried to get my husband but I couldn't get through to him —the lines were busy. I got the school, "If you evacuate my children, where are you going to take them?" "Well, you're allowed to come and pick them up." "Okay, I'll be in to get them." And I thought, "What am I going to do with them? If I bring them home, they'll be closer to TMI." They told me at work to bring the children there.

I went into school and I'm standing in the hall while they call my kids over the intercom. My son comes running down the hall, "Mommy, what's the matter? There are only ten kids left in the room. They keep calling them over the intercom and they don't come back." I said, "There has been an accident." I tried to say it as mild as I could. I was very hysterical on the inside but trying to be very calm and collected on the outside.

I think the thing that impressed me the most that day was how lovely people are. At school, people were all telling us things like, "God bless you." "Do you have a place to go? You can come with me." There was a lot of love there.

MAYOR BOB REID:
Friday was the day. When they said there was a hydrogen bubble and the possibility of an explosion, people went haywire and left town. We estimated that between 30 and 35% of the people left. The schools were in a panic. A lot of the kids thought about dying and wrote their last wills and testaments. *Fifth* and *sixth* grade kids! People were concerned. You could tell they were afraid because a lot of people who left town left their doors wide open, unlocked. They just put anything in the car and took off. They had to run to the bank to get money to go where they were going. It was just a mess.

The news media was all over the place, trying to talk to people. And people were trying to get away from them, trying to get out of town. People didn't understand why these people were *coming* when everyone else was trying to *leave*! The news media would talk to them and people would say, "Well look, I don't have time to talk to you now—I'm trying to get out of here."
Q.: What was it like for you implementing the evacuation of pregnant women and young children?

It wasn't too hard. At that time it wasn't too hard to do anything. People were scared as hell. They didn't put up any argument. They were ready to leave. I had a bus set up in front of City Hall for pregnant women. But heck, most of the people who were in that condition left themselves.

VICKIE DISANTO:
We tried to grab everything that we might possibly need. As many clothes as we could put together, records, birth certificates, my son's baby book and pictures, anything we could think of that

6

couldn't be replaced or that we would need in the immediate future. We were calling everyone we knew, trying to find out where they were. Some people had left town at the same time we did. We had no idea where they were. They had no idea where we were. We were trying to keep in touch at least with relatives and get phone numbers or general areas that they'd be in. We were splitting up in all different directions. We didn't know whether we'd ever see any of the people we had associated with all our lives ever again.

JANE LEE:
It's like your house is on fire and you only have so many minutes to think. You don't know what to grab first and you don't know what to do. You don't know where to go and, if you do go, you have to go with the idea that you're never coming back.

I was in a state of panic, trying to pack the car. I lost things—the keys were stuck in the trunk of the car and I couldn't find them anywhere. It was one hell of a horrible experience to go down that road and look back at your house and know you're never coming back.

And every day Jerry and I came back to milk the cows while that hydrogen bubble was building in that plant.
Q.: How far away did you go?
Seventy miles. One way every day.
Q.: And you came back here?
Every day. We rushed in here, fed those cows and milked them. They only got milked once a day. And we tore right back out again.

BILL WHITTOCK:
People were loading bedclothes and their belongings and their pets and their children into their cars and pulling out. Just leaving, not saying where they were going or anything else. The town looked like High Noon in a Western. The streets were pretty well cleared.

MAYOR BOB REID:
I called for a curfew that Friday. Traveling through the town that night you could see an awful lot of dark homes. It was kind of eery because no one was on the street. You might see one house that was inhabited and then all the houses around it would be dark.

I talked to people. One school nurse said her whole neighborhood was dark. Everyone had left. And she said it just felt *funny,* looking out the windows and seeing all the homes around here dark and not hearing anything, not a car, nothing.

PAT STREET:
I was sitting there Saturday at my in-laws in Maryland trying to decide what to do. I said to my husband, "Well, I know we can't go back and empty the house out, but I would at least like to have clothes because if we don't have clothes we're not even going to be able to get a job." And so we decided to come back and get some clothes. When we were about to leave on Sunday morning, my mother-in-law was crying and just about hysterical. I said, "Mom, we have enough sense if we hear anything, no matter where we are,

Government officials, including White House aide Jody Powell, center, prepare to tour Three Mile Island.

even if we're two miles away from getting clothes, we'll turn around and come back."

On the way down I got really relieved. "President Carter is coming to Three Mile Island," I heard on the news. I felt really terrific. If the President was coming, I wouldn't have to worry. I could pick up my clothes. They're not going to let it blow up while he's there.

When I pulled in the driveway, our neighbors pulled in a few minutes later. I was so glad to see them because it dawned on me while I was away that I might never see anybody I knew again—we might be scattered to the winds. We decided to sit down on the front lawn and wait for the President because he was going to come right down this road—he had to to get there.

"Hey, let's have a party." I had a bottle of rum but no one had any coke so we had to be satisfied with pepsi. And we brought an extra cup for the President. We were sitting on the lawn and he went by and waved to us. I got stoned sober. I was really awed! "Wow, the President! The world's all right, everything's terrific!"

Later I read the NRC transcripts and found out that the time they were in most doubt about the stability of the reactor down here was *when the President was here.* I was just dumbfounded! Here everybody thought he was coming to reassure us that we were all right. I was really angry. While he was down at the plant, we were packing. We followed him out of the plant and down the road where he held a press conference. I don't think we were more than three or four miles away til he was out of the press conference and gone. It was *horrible.* It was like leading lambs to slaughter. It could have been really catastrophic.

DR. MICHAEL GLUCK:
The anguish at the height of the accident was really intense. When we left our house, we had no idea we were evacuating. We took just the clothes on our back and that's it,period. We were debating whether to go back home and maybe salvage a few possessions, should the meltdown occur. We felt when Carter was here might be a safe time to come down and get our stuff, but we decided it wasn't worth it. And we never came back until about thirty days later.

9

PAT STREET:

We stayed in Maryland only until the end of the week because we wanted to get back to work. My husband had taken a week's emergency vacation. I had just called and told them I wasn't coming in to my job. But we wanted the kids to stay in Maryland longer. When we left Maryland, my son said, "Well, if you're going home to die, I want to go with you." *Children aren't supposed to have thoughts like that!* They really aren't. We asked Michael when he thought we should all go home and he said, "Wait until they say it's okay. And then wait a few more days because they don't know what they're doing."

Q.: Your son?

My son.

Q.: And how old?

He was eight at the time. The odd part was we listened to him. We didn't bring the children back the day they lifted the evacuation, Monday. He stayed until the end of the week. He missed some iodine releases because we listened to him. I'm starting to listen to my children a lot more.

Nightmares, Fears and Worries

DR. MICHAEL GLUCK:
The most devastating effect that I have seen on this area has been
severe psychological depression. In my own practice I've seen a
number of patients who have come to me after the accident, telling
me they cried every night; they were unable to sleep; they were
afraid for their children's lives and their children's futures; they
no longer had a secure home. Many of my patients have woken up
screaming from night terrors and dreams about the plant. A lot of
people have had just vague complaints of not feeling well, being
tired all the time, just basically being sick to their stomach. When
I've questioned many of these people, their illnesses basically date
back to the accident.

PAT STREET:
Being honest, I don't feel right about being here.
*Q.: Do you worry about the dosages of radiation you might have
received?*
  Not myself, but for that little one crawling under the table and
the other one that's into reptiles, amphibians and football, yes.
Q.: What kind of thoughts go through your mind about that?
  Really horrible ones. Every time we even have a cold around
here, I get scared. If my children get tired, I get scared. If they
get sick now, I wonder *'is this it?'* I know a lot of people, especially
the people from the NRC, would call me an emotional woman.
But frankly I don't believe them any more. I would love to find
someone I could believe who would tell me it's all right, but I

11

haven't been able to, and I won't be able to. According to the President's Commission on the accident at Three Mile Island, "Our emotional stress was very limited and it's over with." It isn't over with as long as I'm alive.

My son, Michael, wants to know why we can't just attach a little ring to the island and pull it away some place!

VICKIE DISANTO:
Back in September I had the most miserable dreams about running and running. I could always see the towers behind me when I was running. I'd try to go into a house and they'd say, "No, you can't come in here." And I'd run to the next place but everybody would tell me, "No you can't come in." It was like I was running from it because I knew it was dangerous, but there was just no place to go. And I had that dream for three or four nights in a row. The funny thing was I mentioned to a friend of mine that I had been having dreams about TMI. And he said, "I have too." We compared them and they were the same.

On the surface, we have very normal lives here in Middletown, but underneath there's always this presence. It's always with us. I usually think about it at least once a day. There are some nights every now and then when I would just lie in bed and it would strike me and I would cry and cry because I'm worried about my children.

There have been times before when we would have been able to ignore the fact that the towers were down there, but now you can't see them without thinking something. We don't know what to teach our son. We used to go out for bike rides all summer long— he and I—and he'd ask about different things and what they were. And, of course, there were always the towers. I didn't know what to tell him. I don't want to scare him, at two years old. But finally we settled on yucky towers and that's what he calls them now.

We still don't have information. We don't know what's going on from day to day except through rumors. They still have lots of problems down there at the plant because there's so much contaminated material that has to be disposed of. We don't know whether they can do that safely. So, we don't want to live here with all that going on.

Vickie and Albert DiSanto with their children, two year old Albert (T-Bird) and one month old Angela.

I'm just so nervous. Every time I hear something new I think, "Oh, it's starting all over again." Or every time I hear a report on the news about some other place, I think they're going to have to go through the same thing we did. I pity those people. I don't want it to happen to them.

People say to me, "You'll get over it," but this is going to be hanging over me for the rest of my life! I'm hoping people realize that they're going through a day-to-day routine just like me, but underneath mine there's something they might have to deal with some day. And do they want to?

Q.: A constant terror?

Yeah. Do you want to have to deal with that? Do you want to be put in a position where you *have* to deal with it?

My husband and I had a lot of dreams: we thought we were going to stay here only for a while and then move on to something else. But how are we going to do that? The financial investment has a lot to do with our staying here. Moreover, we don't know whether we are going to live to see any of our dreams come true because *we don't know how long we have.*

Any dreams we ever had have been wiped out, even dreams for our kids. We don't know how long we have or how long they have. It's always hanging over us. Things could turn out to be o.k. but they could also turn out not to be o.k. It's hard to have a dream of the future when you have no certainty about anything. There's always gambles you have to take in life, but here the odds are really against us. So it's changed everything psychologically. It's much more depressing. It's harder to get myself up for anything.

Q.: That's quite a burden to carry around.

Yes, it is. It's just something we're stuck with for the rest of our lives.

BILL WHITTOCK:

Q.: What's it been like for you living here over the past year or so since the accident?

Well, there's a feeling of uncertainty. We don't really have any peace of mind. We are uneasy because of this plant being here and because of the possibility of a repeat of this situation. We are not sure that there isn't continued leakage. We have no real assurance.

14

MAYOR BOB REID:

The plant I think will have an effect on the area because it's going to create a controversy all the time. You are going to have people who are upset with the plant but who are not going to leave this area. It's going to have an effect on this town forever as long as the darn thing is there—not only this town, but the entire area. I think people will just sit back and wait for an accident to happen. And any time you have these plants, you might as well sit back and wait for an accident to happen.

Q.: What kind of psychological effect has it had on people?

Kids and mothers—this is where the biggest effect really takes place. Dads are a little tougher. I know girls are thinking right now, "Hey, I want to be a mother. Have I been affected?" They have to worry about this. Mothers are concerned about the child or maybe an unborn child. A woman who wants a family will say, "I don't want to bring a child into this world who is going to be affected either physically or mentally. Will I have a healthy child if I have a child in this area?" Mothers will look at their children and say, "Well, twenty years from now, will that child be healthy? Will he be a healthy young man or a healthy young lady? Will this plant down there have an effect on them now?" Right now we can't tell, but twenty years from now what's it going to be like? Will he or she be able to have children, normal children?

I talked to one particular doctor in town who says he can see regression in some of the kids who come to his office. Kids who were very outgoing before the accident have a tendency now to be a little more quiet and hold onto mom a lot. And the kid who used to climb all over the doctor's chairs and things or read his books, some of the kids now just don't do that. And the doctor wonders if it's the same way at home.

I talked to the NRC people and asked them to bring a child psychologist to town, just to sit down and talk to the kids, not one on one where you lay a kid on a sofa and talk to him, but to a whole group of kids explaining exactly what's going on down there and how it's going to affect them. The NRC people said, "No, it's guys like you who are responsible. If you keep your mouth shut, then the kids will forget it." I said, "Like hell they will."

If Unit 1 should reopen, the kids are going to think about it because they're going to hear about it on the radio or TV. "They're going to open Unit 1, they're going to open Unit 1." Mom and Pop who were raising Cain during the time of the accident are going to be arguing back and forth. The kids are going to think, "Well, boy oh boy, I remember that Friday." And right back it's going to come.

FRAN CAIN:
Q.: Do you ever worry about the dosages of radiation that you might have received living so close to the plant?
Yes I do continually. During the accident I was worried about the animals, I was worried about the child, I was worried about the home—what was going to happen if a meltdown were to occur. I still live in fear. I have nightmares. I have not really slept well. Nerve-wise, all I have to do is look out my dining room window and see Unit 2 of Three Mile Island, directly across the street.

When I started my poodle business, I was very green. It has taken me thirteen years to get where I am now, producing champion dogs. My main objective over the years has been to improve the toy poodle strain. I'm deeply hurt because this has disrupted my life. It has affected me both mentally and physically. Nerves are one thing. I have been ill. Whether it's gall bladder, hernia or what, my body keeps acting up continually, so much so that I cannot operate my grooming shop alone. I've had to call in extra help. I get only a commission out of my own business, believe it or not, because I am not able, physically or mentally, to do the grooming by myself any more.

I can't concentrate on my work. My mind will go back to what is going on over on the island—how truthful they're being to us, whether they're lying some more, what they are doing with the water, if anything is escaping. This goes through my mind continually, night and day.

I don't feel safe. If I leave my house for the day to go shopping or to go into Harrisburg or do something on business, I'm constantly thinking, if there's anything wrong over there, am I going to be able to get back to get my dogs out of here?

16

Trapped

FRAN CAIN:

Q.: How has the accident changed your life?

I feel that I am endangered continually, even now after the accident, as I was at the beginning. I have one daughter in tenth grade who concerns me. I've called realtors in. I've been in touch with Mr. Arnold, Vice-President of Met. Ed. He offered to help and sent a representative in. I told him I definitely want to move. He said they would come back but they haven't done anything.

I have a problem because I have a kennel of twenty-three to twenty-five dogs and have to relocate in an area where it's zoned agricultural. If it weren't for that, if it were just my husband and my daughter and myself, I would pick up and I would be long gone. I would not stay here. The homes that are being sold down here are being sold at a loss. I heard about one man down the street who got $10,000 less than what he was actually asking. I can't afford to do that.

Q.: So you feel like you're in a real bind?

Definitely in a bind. I would like to get out. During the accident, we evacuated the kennel for approximately nine days. I went to the NRC and they told us to move all the breeding stock out because of the proximity to the plant. The bill was over $1,000. I submitted a claim to American Nuclear Insurance.

Q.: Did you receive reimbursement?

No. None at all. I have the letter. It stated that "I *may* have sustained a substantial loss." Well, I definitely did. They have never reimbursed me one dime.

I have a "FOR SALE" sign on my home, but it's very hard for me to relocate. Because of the dogs, I'd have to find land that was zoned agricultural. I really feel we are trapped here. Financially, I just cannot walk out of this house and relocate. It's almost impossible the way mortgage and interest rates are today.

Q.: Have you had a hard time getting offers from people to buy your house?

Yes. Somebody made the remark they'd give me $10,000 for it. (laughs) The homes in this vicinity today are worth about $55,000. I have a business, my grooming shop, that I would have to sell also when I sell the home. I have between $85 and $90,000 tied up here between the home and the business.

Q.: What have you been offered for it?

The only offer I had was for $10,000.

Q.: That was the only one?

Yes.

Q.: And other than that, nobody wants to buy your house?

That's right.

Q.: So what are you going to do?

Wait it out and see what takes place across the road.

VICKIE DISANTO:

We put our house on the market for a while. We had a couple of people come look, but we couldn't sell it. No offers. Besides the obvious reason for moving, we need more space for the baby. We needed money to put down on another house and couldn't afford to take a loss. If we could afford to, we would just leave this house. My husband keeps on threatening to just leave the place, declare bankruptcy and move off somewhere. But no place is really safe, these days.

Our future seems to be pretty much tied up in our house. In our society, things are pretty much based on property, right? When we bought the house, we were either going to move on from here to a bigger house with more land or add on here as the mood struck us. Right now, our house is of no value. We can't sell it and move on. We don't want to add on here, because that would tie us to this property even more.

18

What do we have to live for? We don't know whether we have a purpose any more. If I were someone of sound mind, I would not want to buy this house.

ELIAS CONLEY:
The valuation of our homes has gone down more than half, I believe. They wouldn't be worth half of what they were before the plant opened up.

JANE LEE:
We don't ever want to give up our farm. We love it. We're very unhappy with the fact that the plant was built 3.5 miles from our farm.
Q.: What's it like for you living here?
It's a fear. We live with it every day. We can't plan like we used to. We would like to put in a storage system for manure—that's called a slurry store—so that in the spring when we need it most, we could get the nitrogen out of it and spread it on the fields. A slurry

Assistant manager of a dairy farm, Jane Lee.

19

store with all the equipment costs $45,000. We can't do that because tomorrow we might have to pack up and leave. So we cannot invest in anything that's going to be nailed down, that we can't take with us.

DR. MICHAEL GLUCK:
We have bought a house and we're moving twenty miles away to Carlisle, Pennsylvania. Both physically and emotionally, it's been too intense of an experience living near the power plant. I am committed to the Harrisburg area, however. I have a practice here. I'm committed to my patients and my work.

The advantage to me is that I'm a physician and I have the economic means to leave the area. I think there are a lot of people who really want to leave but can't. It's difficult to sell your house right now and it's difficult to buy a house right now. It's become a very intense situation for a lot of people who really want to make a move but find it economically impossible to do so.

BILL WHITTOCK:
It's awfully hard to just move out quickly without a lot of planning, especially when you have your life's interests in the area and your savings are sunk in property. Why, you just hate to leave it.

We wish that the plant wasn't here. My wife has been asking whether we could move to the west coast. She'd like to go out there. I guess there are nuclear plants out there, too, aren't there? You really can't get away from them, can you?

Suspicion

JANE LEE:

I'm on pretty firm ground when I say that I know, over and above the accident, that we have been exposed to more than our share of radiation because of the total incompetency of the people who are operating that plant. There is no question about it. The Kemeny Commission report, if you read it, proves beyond a shadow of a doubt that the people who were operating the plants down there simply do not know what they're doing.

And the same people are still operating them. They want to bring Unit 1 back on line. During the accident they kept telling us there was nothing to worry about, that everything was under control, and all the time *nothing* was under control. As a matter of fact it's still not under control. Most people don't know we still have an ongoing accident at TMI.

I used to monitor these plants. You can see them from the house and when they'd shut down, I'd call the newspapers, both the York and Harrisburg papers, and say, "Hey, Three Mile Island has shut down. Do you know why?" "No, we don't know why." I'd say, "Will you call Met. Ed. and find out?" It got to be a standard joke because I was calling them so often. Finally one of the guys stopped in and talked to me one day. He said, "How do you know the plants are shut down?" And I said, "Because they emit steam and when they're closed down the steam is not emitting. If they're closed down, there's got to be a reason and I want to know what the reason is." So I believe they weren't even reporting half of the shutdowns to the NRC. The reporter told me that he called

21

Met. Ed. and they said, "No problem. Just some minor thing. We'll have it all back in operation in no time. Don't worry about it." The reporter says to me, "What do we know? We don't know for nothing! You know more about what's going on than we do." And I said, "Well, that's a hell of a way."

Nuclear power plants are always emitting radiation. It isn't a continuous thing, it's sporadic. It's part of the routine operation especially when they begin the operation or shut down. And you never know how much is being emitted. That was in the paper.

Q.: Who said that?

Met. Ed. admits it and the NRC admits it. There's a certain amount of emissions coming from a nuclear power plant.

MAYOR BOB REID:

The thing that frightens me is how many accidents they have been having at these plants that we don't even know anything about. It so happened that this one was serious enough that the news media got hold of it and really took care of things.

People tell me, "Boy, I'd like to get out of here. During the height of the crisis I took my family and left." Then I say to myself, how much radiation have we been exposed to in the five or six years that plant has been in operation? I wonder if it was senseless to get up and leave that day. Five years that thing was running and there was never a report of an accident. And I bet they had *hundreds* of accidents where radiation was released that we don't even know about. I even told people, "You might as well not run. You might as well forget about it. I really think that you were affected over a five year period."

Q.: Did you have any evacuation plan?

We didn't have anything. When I took office in January 1978, I asked about a disaster plan. We didn't have one. I said we should have one. I said, "We have a chemical plant in town. We have the Am-Track lines going right through town. We have the turnpike. We have Route 283. We have the airport. And we have Three Mile Island. We should have something." I gave a talk at a Farmers' Banquet on the subject of emergency planning. After that, people started calling me an alarmist. They actually called *me* an alarmist. Some of the farmers came up and said, "Bob, they're laughing at you in the back room."

Q.: This was before the accident?

Before the accident. I would get criticized if I walked down the street. People would say, "Hey, this guy's an alarmist." I've heard it. At that time I had started to work on an evacuation plan. I even tried to get all communities in this area to send representatives to a meeting. The first meeting was well attended but at the next meeting fewer people showed up. The third meeting I was there by myself with my Civil Defense Director. Then the accident took place.

JANE LEE:

What I'm about to tell you is only the tip of the iceberg. Back in 1973 I began to log incidences of abortions, still births, birth defects and abnormalities in farm animals and livestock. At first I logged just what was happening here on this farm. Then as I began to get reports, I began to find out that it wasn't just this farm that was having problems—it was most of the farms in approximately a five mile radius of TMI.

In the investigation, what I have discovered is that everything is crucial to log. A still birth or an abortion may seem like a normal occurrence to a farmer. A cat that's born with some type of a physical disability, a farmer would usually not give second thoughts to—it wouldn't be very important. However, if you have enough of these occurrences, you begin to get a picture that something is drastically wrong in the environment.

In this valley we are socked in by two mountain ranges, one right close to us and the other one farther down. We get a lot of inversion, a lot of fog and heavy air—the atmosphere is very heavy—so when the wind is traveling this way and anything is released from the plant, it comes right down within a five mile range. It has no other place to go but down.

We have seventy head of cattle, and this includes milk cows, calves, heifers and dry cows. We have four goats and thirteen ducks. Let's see, we've got about twenty cats now, we used to have thirty. The cats and the ducks seem to be the most vulnerable. We have at least three male toms right now that will not breed. They're old enough to breed but, evidently, they're sterile. They won't breed at all. We encounter a lot of miscarriages in the cats. You know, they swell up and look like they're pregnant and then there's nothing. The

23

litters are very small now, anywhere from one to four is the most you'll see in a litter any more. Usually a litter runs at least six. The kittens are very small and the development stage is slow. It's unbelievable. We have kittens out there that were born in the fall and they're still small. You know, they just won't develop. Three kittens in a litter on this farm died after the accident in June. Eight kittens died here in July 1979, five in one litter, three in another.

It's possible that this is occurring because they have a higher reproductive rate. Radiologists agree with this. They say that if you're going to find a problem, in all probability, you will first discover it in cats because of the high reproduction cycle. You know, they live in the environment, hunt in the environment, feed off the environment and they bathe themselves in it. So if there's anything wrong in the environment it's going to show in cats first.

For ducks the same thing is true. Usually in a duck's nest you get about fourteen eggs. Now you never get 100% on those but in one nesting last spring after the accident, we got one nest we didn't get anything out of. Out of another nest we got one. And in another one we got four. And out of those four, one was striking its foot and its whole digestive system seemed all fouled up. It couldn't eat. It just blew up. It was the most horrible thing you ever saw.

The summer before the accident we had one duck which was odd. This duck never developed properly. It got so it grew just so far and then stopped. The feathers when they came in stood straight out like quills. It also had something wrong with its leg—it couldn't walk. When you would go to pick it up, there wasn't any flesh. It was just bone. Just hard, brittle bone. Anyhow, one day I went up and it was dead.

Emma Whitehill lives right down the road from us. She's been on the farm since she was twelve. I'd say she's getting close to eighty now. In 1978 her ducks laid 290 eggs without a single hatching. 290 eggs!

Dr. William Bush is a medical doctor who has a horse farm up here northwest of us. That's where the largest amounts of plumes went. He shipped four or five thoroughbred mares down to Kentucky to have them bred. It costs $3,500 a horse to have horses bred there. They were confirmed pregnant in Kentucky and he brought them back. Before the local vet could check them, two had aborted. The vet confirmed to Dr. Bush that the two mares had aborted. One of the others had a problem at birth because the foal got lodged in the channel. Another one was a still birth. And another one was still born. He lost every last one of those foals. He said to the vet, "Now what in the hell am I going to do?" And the vet said, "Well, the only thing I can tell you to do is move." This was *before* the accident. A lot of what I'm telling you happened before the accident.

There are two things wrong with the pigs. Their heat cycles are way off. But there's something else that the vet, Dr. Weber, told me and I have this in writing. He says that he usually delivers one pig a year Caesarian. That's normal. He's now encountering two Caesarian deliveries a week. We're also seeing a lot of cysts on the pigs' ovaries.

The radiation is striking at the reproductive system, the area most vulnerable to radiation.

The large animals—the steers—have suffered blindness and multiple fractures especially in the rib cage. In fact, when one animal was butchered, I talked to the butcher myself. He said, "I never saw bones like I saw on that animal. The ribs, I just cut those right through with a knife.

Q.: And usually?

Now a steer rib you'd have to cut through with a saw. Like when you're cutting short rib or something, you have to take it through a power blade.

Q.: They're brittle?

No, just soft. Really soft. That's why their bones are fracturing. That's why these steers slip and fall. One totally crushed its hip. They had to destroy the animal.

Q.: How far were these animals from the plant?

Within a five mile radius. These are all people who live in very close proximity, right along line with the plant.

DR. ROBERT WEBER:

Q.: What problems in farm animals have you witnessed in the past few years here in the vicinity of TMI?

25

Dr. Robert Weber, veterinarian, with friend.

Feeder calves were the first animals we had problems with. They weighed between 250 and 450 pounds. There was no trouble with these calves at the dairy farms, but when the dairy farmers sold them to their neighbors, there were problems like multiple fractures, blindness, chronic arthritis, anemia and what looked like starvation and deficiency.

We took samples from these animals and some of the animals themselves to the laboratory. We found multiple fractures in all bones of the body. These would not heal. Even when we had a calf with just one broken leg, it would not heal.

We've seen this problem for about two and a half years before the accident. The last case was in June, three months after the accident. Since the plant has been closed, we haven't run into this problem.

Q.: How many calves were affected?

I would say at least 50% of the calves in this area. Some people lost all their calves. I suggested that the owners of these calves feed their cattle anywhere from three to five times the normal requirement of vitamins and minerals.

Q.: Is this recommendation unusual?

If there is a severe deficiency, such as what I saw here, this type of treatment is necessary.

Q.: Would you say that the area, for some reason, is becoming deficient in vitamins and minerals?

I think something somewhere was making the required minerals unusable to the animals and that therefore they came down with these symptoms. But because I haven't run any tests on the soil, I don't know whether that's even true. I asked the state to do it and I asked the farmers here to do it on their own. But nobody did it. So we'll never know, I guess.

Q.: As a veterinarian, does it alarm you to see these problems?

Sure. When I see these plants in operation and then I see these young animals apparently normal one day, then the next day unable to get up because of broken bones, of course it's alarming.

CHARLES CONLEY:

I had one steer that got down with a broken leg so we killed him. Another one went down about the beginning of February that winter. I got the vet'nary up here. He said I got to put in more vitamins and minerals or I would lose the other calf I had.

So I kept feeding that steer. He'd hobble around and try to get up on his hind feet but he couldn't walk because he had thick knees on the front. He'd go in circles, down on his knees, back and forth through the stable. I fed him all this summer til the fall. When I went over there one Saturday morning I saw him laying there. I heard him breathing heavy and his neck was stretched out. I guess his heart conked out. That was the end of him.

Here's another thing. Toward the end of January we had the biggest snow we had all winter. Then it turned around for about a week and then we got three inches of rain—it rained that much in my watering trough. I let them big steers out and they drank that water. It give them the diaree. If you would have been standing behind them, you would have got a face of it. So that put 'em off their feet for about a week. So every time it rained after that I cleaned the watering trough out, because when it rains, that fallout comes down with the water. My nephew out here was cleaning out his watering trough every time it rained. I didn't know it. He said it was in that rain every time it came down. See, every time it rains the winds come across from the plant and it's a down pressure and it comes down with the rain. We caught water here at the porch, it would be just like white milk. All milky on top. You couldn't use that water. If the animals drank it, why, they'd just lie down and get sick. Well I talked to some people about it and they said that comes from all these airplanes flying around in the air. But as soon as them plants closed down you didn't see that on that water any more.

VANCE FISHER:
We've had cattle trouble here for the last four years. In fact, I've lost five, I think, because of the TMI situation. The problem seems to be that they become paralyzed and usually lie around for a period of at least six weeks, sometimes up to two and a half months. I've had them get up after two months and I've had them die too. We should have had at least eight, possibly ten kids from the goats last year. We had one that had a miscarriage. She had three but they were all born dead. We had, I guess, four that didn't breed. Other times when the goats were bred we always had kids, usually two and on some occasions three. My family has been here for two hundred years, right here in the same area. We never had such problems with cattle 'til several years after TMI came.

I've complained to quite a number of state people, people who I thought should have an interest and responsibility in it. But they have completely ignored us. I made half a dozen calls to the Department of Environmental Resources trying to find out what pollution we were getting from the towers. They insisted they didn't know. In fact, they insisted I should talk to Water Pollution. They transferred the call to Water Pollution so I asked what was in the river. They insisted they didn't know. But they insisted that I should talk to Radiation Control so they switched the call to Radiation Control. They asked who I was and when I told them, they said, "We don't talk to common ordinary people."

ELIAS CONLEY:
I've been farming all my life. I have one hundred and twenty-three acres. Corn, oats, wheat and hay is my crop.

Q.: How were your cattle affected?

Well, since that plant opened up I had some that went up to five hundred pounds and then got a broken leg. I don't know what caused it. They got down and died finally. Now you tell me what causes it. I never had one before with a broken leg or broken bones. It's something that causes the bones to get weak. The vet told us it was a lack of minerals and vitamins. His suggestion was that we double or triple the vitamin and mineral supply. Why should we have to go and buy so many vitamins and minerals? We didn't before the plant opened up. They're expensive.

GLADYS CONLEY:
If there's a problem, we feel that it must be something that settles on the ground. We can't help but feel that there must be some connection with the plant because we never had trouble before.

CHARLES CONLEY:
The hogs—it affects their breeding. I had four sow pigs last fall. Well, I got them pigs last summer and they never showed any signs of coming in to breed. They should come in when they get about four months old.

Q.: You mean they didn't come in to heat?

Yep, they didn't come in to heat to breed. I got two sows this winter out there. They come in to heat a couple times already this year. But the plant ain't in operation now. But if they start it up it'll do the same thing again.

Q.: Do you think the problems you've had are caused by the nuclear plant?

We didn't have it before. It has to be from that. I was hit hard here because when it rains it comes right in across here. It runs all the way in this here circle from the mountain, the whole way around here clear down to the Conawoga Creek they have trouble.

The fallout affects the animals. It affects the breeding. They have trouble getting their cattle bred, you know. Some of 'em, they artificial breed a lot now and then they decide to keep a regular bull and breed the old way. Well they still had trouble to get 'em bred. And then, some of 'em have the calves born dead.

A fellow I know with goats—the goats didn't even want to breed. A woman who lived on the other side of the river above the plant had several goats—she said she couldn't get 'em to breed.

My nephew who lives back here had rabbits. The kids had thirty little rabbits right before that accident and it killed 'em all.

Q.: When did they die?

Right after the accident. That gas in the air affects them.

DR. ROBERT WEBER:

Right after the accident, in fact while the accident was going on and even slightly before, we had problems with pigs. We had to do a lot of Caesarian sections on many sows. Ordinarily, I don't do more than one Caesarian a year. In this particular area, I was doing one a week. Then during the lambing and kidding season of this year (1980) we had an untold number of Caesarian sections. Usually we do them rarely, maybe one a year but this year we had two of them a week. With such a few goats and sheep in the area, I thought this number of Caesarians was exorbitant. The animals that got here in time to have the Caesarian section lived, but the ones they let go to see whether they were going to be able to open up on their own accord, why, they were born dead.

Q.: So what was the problem? Why did all these animals need Caesarian sections in order to give birth?

The hormonal makeup of the animal was not working right so that when they were due to deliver their young, they didn't open up, they didn't dilate.

After the lambing season, we had a number of cows that didn't dilate. So we gave them injections of hormones. In the cows, the injections seemed to work, but in the sheep and goats it did not work. It didn't matter how many days I waited after I gave the injection, it just didn't work. The cows, within five days to two weeks after they received the injection, delivered normally. And that's the way it should be.

Q.: How far did these animals live from the plant?

Mostly within five miles.

Q.: Do you know about other reproductive problems that animals in the area have been having?

In the last few years we've been having quite a problem breeding horses, but what the difficulty is I don't know. We're just unable to get them bred.

After the accident, there were very few baby kittens born. On some farms for the whole year afterwards, they did not have any

30

kittens until spring. They didn't breed all that time, from last March to this March. Now there were some places where the cats did get pregnant and had kittens, but they did not live. They were born dead or died soon after birth or if they survived the first few days, they didn't grow and died later on.

Q.: Would you say that the incidence of the problems are high in the area near the plant.

Why, sure. I practice in parts of three or four counties and I don't have these particular problems anywhere else. I've been working down there for thirty-three years and I know something is wrong.

Q.: Do you feel the problems may be caused by the nuclear plant?

Well, it's like a big monster over there. I know there has been a problem in the area for at least two years before the accident, but I don't know what the cause is. These problems weren't here before the plant was in operation—they occurred while the plant was in operation. And since the plant has been shut down, a lot of the problems aren't here any more. So this might indicate that there's something wrong with the plant.

Farmland near Three Mile Island.

FRAN CAIN:
I'm afraid to let my dogs go outside. They chew on rocks. If anything has fallen out, my animals could be affected by this.

Q.: Have you had any problems with your puppies?

The only thing that we have had is one blind puppy. There are no sockets in the eye.

Q.: No sockets in the eye?

There are *no* sockets whatsoever. I've had abnormal bleeding in my female dogs over the past year during their heat cycles. My toy poodles bled drippings up to the size of twenty-five cent pieces or more which is very unusual for a poodle this size. I completely stopped breeding after the accident because I was afraid to breed.

I've taken a loss through puppy sales. I've also taken a loss in breeding champion studs because people just don't feel like shipping them into this area for breeding. Some of my grooming business is slowing down because a lot of people, even customers of mine that have come here for years, don't even like to come down here and see the plant as a reminder, even though they live around the Middletown area. We are never out of sight of it. It's here night and day.

JANE LEE:
On our farm in 1977: January 4, calf born; died January 11. February 3, beefalo calf died. August 9, calf aborted. September 6, calf died, three months old. September 9, calf still birth. October 5, cow slaughtered: diagnosis, cancer, confirmed by vet. December 15, goat aborted. Other goats did not conceive. This is all 1977. In 1978, a day old calf died in May. Four female ducks laid seventy eggs and out of seventy eggs there were seven ducklings and one was a mutation.

This is an update of the history of our farm. I wrote the animals' names down. The first one is Alice. Alice could not breed so we sold her, 10/30/79. Edith had a calf due October '79, she aborted and we sold her, 8/9/79. And Babe had a calf in July '79. She came in heat, off and on for five months. She was examined by the vet three times and on 12/30/79 a cyst was discovered and we sold her January '80. Baby Bunting—still born 12/16/79. Barb calf 10/14/79 came in heat, breeding negative. Treated for cyst by vet

32

12/31/79. Dorothy had a stillborn 5/79. Orange Juice—don't you love these names?—Orange Juice was bred early October '79, negative. Grace checked by vet 10/11/79, breeding unsuccessful, has cyst. Maggie ditto.

Louise Hardison, who lives in Middletown, has lambs and goats. This is her living. Three mother rabbits, one bred twice, delivered eight rabbits per litter. All baby rabbits died. One female goat died. She reached full term for delivery carrying four kids. Imagine, all dead. After the accident one pregnant female sheep died and three lambs two to three months old died. Louise had one sheep that I know of that was born with one eye. I know eyes are affected by radiation.

James Fitzgerald is an agricultural teacher. In 1978 he had two steers who were born blind and with soft bones. Both had to be destroyed. In January 1979, one steer weighing four hundred pounds, despite confinement, broke both hips. That animal was confined and both hips broke! Now here's four steers that died, February 26, March 10, April 8, and April 18.

Q.: What is the percentage? Is it very high?

Yes. It's very high. He lives very close to the plant.

Mr. and Mrs. John Kaufman live across the way. December 1, 1978, a steer weighing two hundred pounds lost control of its hindquarters. January 1, 1979, a steer weighing three hundred pounds lost control of its hindquarters. Both had to be destroyed. He found cats which were aborted, four of them in different stages of pregnancy. Some of the kittens were dead, some were dying—all littered over his lawn. They all died. The Kaufmans are right on line with the plant.

Mary Ann Fisher lives in Middletown. In 1978 they had a litter of kittens, three weeks old, that died overnight. They had twelve geese that laid one hundred eggs. Results: one hatching, all of which died. In 1979, January, four litters of cats aborted. Four litters! One full term litter stillborn. Four heifers unable to conceive thus far. Geese laid eggs again. Results: nothing.

Did you hear about the Claire Hoover farm? Well, this was on national television. In April 1979, after the accident, four of his cows died all in the same month. In fact, three of these died on the same day. Now that's a pretty high percentage. One cow was down sick. Another one died. Cow aborted, both cow and calf died. Cow died, discovered twin fetus. The two calves died, also.

Joe Conley's farm is very, very close to the plant. He moved all his stock and his family the day after the accident and did not return. He suffered a high rate of reproduction problems. His wife had two miscarriages. Also before they moved, Mr. Conley and his son, who's eleven years old, developed problems with their legs. Their veterinary bill absolutely soared way out of sight. He said that his cattle suffered from nervous disorders, so much so that he had to hobble their legs to milk them, they were so high strung. Did you ever hear of that?

On Charlie Conley's farm, in September 1978, one steer was down, unable to get up, and it died. December 1978, steer was down, unable to get up, and it died also. His vet bills were over a hundred dollars a month. The vet informed him that he would have to feed two and a half times the required amount of minerals in order for the animals to survive. No farmer can stay in business if he has to spend over a thousand dollars a year on minerals alone.

Last summer ten kittens from two separate litters all died. This was on the Conley farm.

I don't use information unless I actually get the person's signature. All these reports I have were signed. I didn't want anybody backing away after they had made a statement. I wanted their signature on it. They read it and then they signed it. When people tell me stories, if they cannot give me any concrete proof, forget it, I'm not interested. I don't want hearsay. I want only what I can prove.

The evidence to me is so overwhelming. This may seem incredible, what I have told you. Remember it's just what I myself have done. There's a lot of things going on that I haven't even logged because I don't have the time or the money or the energy to go and see all these people, with the cost of gas and everything. And you should do a follow up all the time on these people.

I would say that the number of incidents that we are encountering is very high. I can't give you statistics because I would have to have more information than I have now. When you're dealing with any kind of a health survey, it has to be very widespread. But, see, I'm just one person. I'm the only person who has done anything.

Farmer Vance Fisher.

35

I had occasion to talk at a dinner with a geneticist, a pathologist and an epizologist. In talking to them, I began to realize the complexities of this type of reporting. How in the world do we ever hope to prove that what is happening to the animals is from radiation? We have no studies at all that we can go on as a guideline to make a comparison before the plants and after.

One of the doctors said to me, "Look, maybe you can't prove that the problems are caused by radiation, but the only way you are going to be able to force a study on the environment is to log what is occurring and keep on reporting it until finally they have to come to terms with it and recognize that there is a problem. This is how they got the study done on DDT."

There's an awful lot going on in the animals that we don't know about that has not been logged. There's been a great deal of effort by some farmers to conceal the evidence because they feel threatened in their very survival. And, of course, the reimbursement by the government will be practically negligible as far as what they will get in the value of their land, their animals and their business. Their whole life is tied up in the land. It's a very difficult thing to go talk to farmers. I have encountered strong resistance and have had to return sometimes as often as three times before I could even begin to get them to open up.

I believe that part of what is happening is radiation. But I also think there is another factor here that is just as serious as radiation.

VANCE FISHER:
After doing a little study and research, I came to the conclusion it was white muscle disease, which is caused by a lack of selenium. All the towns up river all the way through Pennsylvania and way up to New York, put a lot of chlorine in their drinking water and a lot more in the sewage when they put it back in the river. Selenium has a great affinity for chlorine.

When we get what we call an up-river wind or a wind from the east, it blows into this valley. Anything blown into this area just seems to linger. So any pollution blown into this valley would settle to the ground. All that chlorine that came out through the towers as steam from the river tied up our selenium. If I remember rightly, this makes selenium-oxy-chloride which the cattle can't digest. This area is known to be a little scarce in selenium anyhow but it never gave us any problems until after the towers had been operating for several years. For years I kept paying the veterinarian and the animals usually died. Since I've been feeding my animals selenium I haven't lost a one from white muscle disease. And I've actually saved four that were paralyzed.

JANE LEE:
For two weeks just after the accident, there wasn't a sign of a bird anywhere.
Q.: Anywhere?
Anywhere. When the farmers go out to plow their fields, the birds usually hover all over the fields. It was the weirdest thing to plow the ground without a sign of a bird. There wasn't a sign of a bird! York county, where I live, is notorious for its starlings, but this year the starlings never showed up. Now, starlings don't just come like average birds. They come by the hundreds of thousands in wave after wave and cloud after cloud. In fact, we were saying that something was going to have to be done about the starlings because they were so bad. But they never showed up this year.

VANCE FISHER:
This spring, after the accident, when I was plowing in the fields, for days we just didn't see a bird. Usually, we have all kinds of birds out there.

CHARLES CONLEY:
I found several robins and starlings laying around on top of the hay in my barn. They flew in there and died. Up at my brother's, the robins flew in the porch and fell dead in a peach basket.
It killed our pheasants. I've seen only one or two pheasants all summer long.
Q.: How many do you normally see?
Well, they pull the corn out, fifteen stalks in a row. I even sowed a lot of corn for them to eat. They just wasn't here last year.
It killed the quail too, no quail last summer.
It killed the snakes that were out there crawling around. I didn't see a snake on the whole farm last summer. I worked all around

over the farm, hauling hay and straw. Husked the corn all by hand and never saw a snake.

It killed the hoptoads. As many years as we've been here there's always been a couple of hoptoads in at the house. In the evening in the summer if you're sitting there on the porch when it gets on towards night, they'll come out there and catch flies, but they ain't there this year

DR. ROBERT WEBER:
For a couple of years I heard hunters complaining that there was no small game. Previous to the accident, there was no small game between twelve to fourteen miles from the plant.

PAT STREET:
An odd thing this year is that we didn't have any rabbits like we used to. You'd come up the driveway and there would be three, four, five or six scurrying to get away from our headlights at night. It wasn't like that this year. I saw only one rabbit this summer. Also, our garden wasn't bothered by animals. Normally you put green beans in and they eat the leaves off down to the stems. Ground hogs get into everything, but the garden wasn't bothered this year. I've had quite a few people comment about this.

JANE LEE:
On the Conley farm, facing the plant, is a big tree. Right after the accident, I guess about two weeks later, I looked up and I said to my sister, "My God, look at that tree. What in the world happened to that tree?" It was all green on the outside but smack through the middle there was a hole just like somebody took a blow torch to it, just as smooth as could be, a great big hole about fifteen by fifteen feet, right through the middle of that tree. It was the damnedest thing you ever saw. No leaves. It was as if something passed right through there. All their pear trees, all their apple trees were defoliated. Their pine trees looked like they were dying.

Oh my, the defoliation was terrible! We never saw such weird things as we saw this year. There was a great deal of discussion here about the defoliation of the trees after the accident. I'd say there was defoliation of the trees for about a ten mile radius from the plant. In some areas the trees were totally defoliated. We do not know what the damage is, except that we know it's coming from the atmosphere. According to a botanist who looked at this stuff, it is not damage from disease. It is something that is coming down in the environment.

These leaves looked like fall leaves, but they were found in July. They were all shriveled up, blotchy and full of holes. The leaves look almost like tissue paper.

We had two good pear trees back here on our farm that always produced very heavily but this year in July they were almost completely defoliated. The fruit remained, which was interesting. The fruit did not drop; it stayed but it was deformed. And it was no good even for canning. A lot of trees right along the river were damaged. It's consistent with all of the reports we've gotten.

CHARLES CONLEY:
Q.: How did it affect your trees?
It killed that big leaf maple over there. It was nice and green last spring when the leaves first started to come on it but later in the summer, you could see the leaves dying on it, and toward the fall, the bark was getting loose on it. It looks in pretty bad shape now. I think it will get some leaves this year but I think it will eventually die.

It killed all the grass the whole length of the garage where water runs off the roof; it killed the grass up at my brother's where it ran out of both rain spouts. I was up to a neighbor this spring. He showed me out his back window how his grass was dying. If the plant was to be in operation for ten to fifteen years, you couldn't grow hay in this area.
Q.: Do you have a problem with your fruit trees?
✕ I have a pear tree that I usually get twenty bushels pears off of, but last fall, on this side of the tree facing the plant, you could see where something went in and hit the tree. The leaves were getting brown and the pears on that side of the tree were no bigger than a fifty-cent piece. I just chopped them up with my shovel and fed 'em to the hogs.

39

Farmer Charles Conley.

My nephew, Samuel Conley, has some young cherry trees. He said it killed them. He's right in this same line from the plant.

It's killing some of my grapes and some of my brother's grapes. My nephew says it's killing his grapes too.

ELIAS CONLEY:

We had pears which got cracked all around and fell off the trees. They were deformed. This was after the accident. I don't know what caused it. Since that plant opened up down there, why my wheat seems to be a poorer crop. I don't know what caused it, I have pretty good crops mostly. It's like a basin in here. Over there is a mountain and up there is another and down below is another one. When the east wind comes, any gases fall out. I believe it's the gases and stuff that's killing our vitamins and minerals in our ground.

VANCE FISHER:

I believe that there's a possibility that we're losing some performance in our crops because of the pollution. There's all kinds of minerals and elements in the soil we need for growing the crops. It could be that we're losing some of them because of TMI. I've seen it when water runs out a rainspout across the lawn. For perhaps twenty feet out, a strip of grass about two and a half feet wide is all brown. I never saw it before this spring. But even under the barbed-wire or the wire fences, there's a strip, say three inches wide, where the grass is dead. And there's no explanation for what happened to some of the trees I've seen.

Q.: What did they look like to you?

Well, the trees died, that's all we could see. We couldn't see any insects. Generally it was on the side toward TMI. So, it's just a process of elimination. A process of suspicion. If the fellows in the State Agriculture Department would get around to make studies of that—the guys who have doctor's degrees and all the years of training and experience—it's quite possible they'd come up with some answers. But if they did, I don't think they would tell us.

Q.: What message would you like to send to other farmers who live near nuclear plants?

Keep an eye out and learn all you can. Watch everything and start screaming when you see anything suspicious. But probably nobody will come to help.

JANE LEE:

When you try to prove that what you are encountering is radiation, they are just going to laugh at you, because you can't prove it, first of all. Second of all, radiation takes a long time to have an impact, except when you have a high reproduction rate. Nobody is going to pay any attention to you as long as the trouble's with the animals. People are going to say, "Well, it's happening to animals. It's not happening to me." But what they don't realize is that animals have a higher reproductive rate. They live in the environment and they feed off the environment.

VANCE FISHER:

My question is, if it's doing this to our animals, *what is it doing to the people?* We raise a lot of vegetables, can some, put some in the deep freeze. We live entirely on meat raised in this area. There are a lot of people, particularly the older people, who eat a very large part of their food which is grown in this area. We have people who have their wheat ground for wheat flour. We have people that have their corn ground so that they can make their muffins and pancakes. And of course that is from grain grown right here in the area. So we don't know how deficient we might be. We don't know when the deficiency will start showing up or how it will show up. There's all kinds of minerals and elements we need for growing crops. It could be that we're losing some of them because of TMI.

JANE LEE:
The things that are being grown in this area are going to be devoid of the necessary vitamins and minerals. The wheat and the corn or anything else that they're growing is going to be sold on the open market. People are going to be eating it and they'll be satisfied as far as their hunger is concerned but their bodies are going to be starving. And we're going to have all these horrendous problems and wonder what in the hell is wrong. And if it continues over a long enough period of time, we will not be able to grow anything on this ground. We don't know how much longer we are going to be able to farm this land if they continue to operate that plant down there. If the emissions that are coming from the plant are going to destroy the environment, then we're not going to be able to stay here. And we're not going to be reimbursed from the government. It's a very, very uncertain type of an existence.

DR. MICHAEL GLUCK:
I truly believe that there will be an increase in the medical consequences in the area from the accident, such as miscarriages and increased incidences of cancer and leukemia. You have to prove it scientifically but scientific studies have not been done up to the present, unfortunately. There is no data on the health of the people surrounding the island before the power plant was built, or what the incidents of miscarriages were before the accident happened, etc. I'm sure this type of study was not done because of the feeling of the industry. You know, they're not going to go out and find data to prove that their plants are increasing the deaths among area residents.
Q.: How accessible are hospital records for a citizen's investigation of health effects?

Farmer Elias Conley.

Hospital records are always confidential and extremely difficult to obtain. That's why it's been really hard to gather statistical data around these plants. It's almost been impossible. There are few medical people who are actively pursuing the problem. I would say merely a handful of us, maybe five to ten doctors, have made themselves visible trying to educate the public on the dangers of nuclear power. As for gathering statistics and studying them, it's really beyond the scope of the practitioner at this point in this area. It requires money and time and personnel, all three of which I don't have.

Someone is going to have to do it. The state right now is planning a study on the health effects of Three Mile Island. Gordon McCloud, who's the Secretary of Health for Pennsylvania, was to be in charge of this study. It was to be an ongoing study for twenty years, surveying the people who live within a five mile radius of the plant, especially the high risk people, pregnant women and pre-school children, and following them for a period of twenty years. However, after actively pursuing his stand on Three Mile Island, McCloud was fired by Governor Thornburgh and replaced. I don't know McCloud's feelings on it but he believes that his stance on Three Mile Island—his desire to pursue this to its ultimate end—has cost him his job.

The people have to be studied for a period of twenty years before you really see what the accident has done to them. You're really not going to see anything before then, unless they pursue still-birth statistics. A study like this, of course, will require money and personnel every year and five years down the line it could easily be discarded.

JANE LEE:
We have a housing development near here which sits up real high. When those plants were operating, many times I could see the steam go over the top and settle right up there on that development. There are pockets of women up there who are having miscarriages. There's been two suits filed already in the area against Met. Ed. The women had still borns. We don't know yet how far-ranging this is because in the Hershey Medical Center the cancer department is so secret that nobody can go into there and get the statistics.

44

PAT STREET:

One person I know personally was pregnant on March 28 and had a miscarriage on the following Monday. There was a person in our group who has a friend who was pregnant at the time of the accident. She didn't get back to a doctor until several months later and found out that her child—the fetus—had been dead for two months. I was told that at the York Hospital across the river there were four still-births the week following the accident, which set a record. I've heard other stories which may or may not be true but they were from people whom I know very well and consider reliable.

I did have a neighbor who lived down here below me whose husband worked over at the island in the control room. I got a letter from his wife right before Christmas. She wrote, "I guess you're wondering what happened to the baby I was expecting. Well, two weeks before it was due to be delivered, the doctor told me it had died for unknown reasons."

JANE LEE:

A lot of people here are developing pneumonia. I'm really surprised at how many. These days, you don't hear of pneumonia like you used to. This is part and parcel of radiation, it destroys the immunity system.

MAYOR BOB REID:

Is this the start of a disaster for the next hundred years as far as my family is concerned? You've got to look at that point, and people are not—they're looking at today only. What about twenty years, forty years, sixty years, eighty years, one hundred years from now, or even five years from now?

PAT STREET:

My neighbor, Vickie, got pregnant after the accident. She was really quite upset when she found out she was pregnant. I went in to see her one morning and she said, "I finally discovered a good reason for being pregnant." And I said, "What?" Her son's name is T-Bird, that's a nickname. She said, "If T-Bird gets leukemia, I'll have a sibling available for a bone marrow transplant." This is what this is doing to people! She isn't a woman who hates kids. If you'd go

45

down there and see her with her little boy, you'd know she loves him. And, under normal circumstances, she would have been ecstatic to have another baby. But this was how it hit her. I felt like I'd been punched in the stomach.

VICKIE DISANTO:
When I found out I was pregnant, it was a real blow because we were still trying to justify having another child under the circumstances. I was really excited about my first child. While I was carrying him in the womb, I had already developed a relationship with him. But with my daughter, born May 6, 1980, it was not so easy. It was like, "I am pregnant," and that was it. Never anything in terms of baby—boy, girl—wondering what he or she would look like or anything like that. It was more like, "If I have a baby, I have a baby. If I don't, I don't." It's harder to look forward to anything any more.

Two women just delivered in the past week. We had talked about it at different times. They would say, "I don't want to talk about it. I don't want to think about it any more than I have to. But there damned well better not be anything wrong with my kid!"

My son is showing every sign of being an extremely gifted child. And to think that something that happened when he was eighteen months old may obscure all those gifts really upsets me. It seems like such a waste. When we should be enjoying our child and encouraging him, we have to ask ourselves the questions, what are we encouraging him for, what kind of life does he have, if anything? I don't think it's fair to him and it's not fair to us. I'm angry because it seems like such a waste that something that happened then could affect him for the rest of his life, when I've tried so hard to keep him healthy and to give him every advantage I can.

JANE LEE:
We live with a terrible fear that David, who is eight years old, may die from leukemia. We have to live with that every single day. I brought four children into the world. When they bring that child and put that child in your arms, the first thing you do is unwrap that child and make sure everything is together. There's no way on God's earth I'd have myself sterilized. But I'd never bring

another child into this world if I was young. No way. Not with what I know.

BILL WHITTOCK:
The radiation just hung right here on the first day of the accident. There was a temperature inversion that morning. Then after the air began to move and the radioactivity began to move with it, they commenced monitoring. So they didn't measure the real heavy burst that was the first release. We don't know what we got. The equipment that they had here down along the creek, what they call a "sniffer," had been through two floods. So that thing was completely out of shape—the gate was hanging open with the lock off the door. That was supposed to have been the detector on this side of the river. It wasn't running at all. It was completely shot.

FRAN CAIN:
We will never know until years to come how much radiation we really got. We will never know because they will never tell us the truth. I don't think they know themselves because when all this happened the equipment wasn't here.
Q.: That must not give you a very easy feeling for the future.
  No, it doesn't.

BILL WHITTOCK:
Q: Do you ever wonder what kind of dosage you received?
Yeah, I do. I'd be quite curious to know what degree of radioactivity I was exposed to. It was a close call. I did taste the metal. They claim that's an indication of a high release. And I understand that there were people over in Middletown who had that metallic taste too.

FRAN CAIN:
We had very bad tastes in our mouths, like an iron or metal taste. It came right in the house to us. And the first time or two I noticed it, the girl in the grooming shop noticed it, too. One day, my daughter was eating breakfast and she said, "Mommy, I have a terrible taste in my mouth." We had it three or four times. This was really,

I think, the iodine. Probably it was the molecules that had been broken up (that we were getting the taste from).

JANE LEE:
We had a metallic taste in our mouths. It tasted like, you know, like when you're a kid and you put money in your mouth? That's what it tasted like. And we all had it.

Jerry had inflamed eyes. A lot of the people had inflammation in their eyes, right around the edge. Their eyes would water and that part of their eye was very, very red.
Q: This was after the accident?
Yes, after the accident, actually while it was still going on. They say that radiation wipes out the thyroid very rapidly. I have had a throat condition ever since the accident.

VANCE FISHER:
I got kind of scorched the first day. I didn't know what was going on and I had outdoor work to do so I was out most of the day. Got a little burn out of it.
Q: Radiation?
Well, it wasn't sunburn anyhow. My face got red. The next day then I heard there was radiation around and I stayed in more than half the day but I did have things that seemed necessary to do outside, so I was out quite a bit of the day.

BILL WHITTOCK:
The NRC man told me that they are still putting small amounts--he classified it as "allowable amounts"--into the Susquehanna river. They are also making "allowable" radioactive releases into the air. They admit that.

JANE LEE:
Every day I pick up the newspaper, it's something new. They release radioactive tritium into the river. I don't know exactly how many hundreds of thousands of gallons of water they run through a filter. But you can't filter tritium because tritium is a liquid. It's not solid, so when they run it through the filter, the tritium goes right through. They dumped all this water with the tritium in it into the drinking

48

Civil Engineer Bill Whittock.

49

water of the communities down below. So Lancaster and Columbia and all those communities are drinking this water with this tritium in it.

Radiation is like Russian roulette. It can affect me and never touch you—you can walk away from it scot-free. Scientists don't know why. But low-level radiation is more dangerous than high-level radiation. High radiation will kill the cells outright, whereas low-level radiation will kill perhaps most of the cells, but leave some intact. Those few are just like the other cells but they will start multiplying so rapidly that before you know it you have a cell in your body that's growing wild. It doesn't take long. It takes only one cell to start the chain reaction of cancer in the body. And that's the danger in radiation, among other things. It weakens the whole system.

Many men are not concerned. They do not consider themselves in danger when in fact they're in greater danger than women. The male organs are exposed and are much more vulnerable to radiation. But you just try to get that across and they think you don't know what you're talking about. They thought I was the nuclear nut for years until the accident and then the phone darn near fell off the wall. "What shall I do? What shall I do?" I said, "It's too late to do anything now."

It's affecting all of us. The radioactive materials are going from the waste into our environment. They're coming back to us through the food chain. So it doesn't make any difference how close you live or how far away you live from a nuclear plant. We're all going to be affected. And in 50 years, if we continue with the operation of nuclear power plants, the people in this country are not going to be strong enough to go out and work on a job, much less defend their nation. And that is a fact. We are not going to be stong enough because it is going to destroy the immunity system. And it will affect all of us. Every last one of us, I don't care where we live.

It isn't going to go away by pretending it doesn't exist. And the longer it's postponed the greater the danger is. Now that's my feeling. Somebody has to speak out. I went for three years with people making fun of me, telling me I was crazy and I didn't know what I was talking about. Now we all live in fear. All of us. I'm not alone any more.

Anger

JANE LEE:

It is "*criminal* negligence!" These people should have been arrested for criminal negligence and there they sit. They are still operating that plant. There's only one word that keeps going through my head. What the United States government has done to its people is a terrible *betrayal.* That's the only word I can use and I use it over and over because that's what it boils down to. We have lost our country. We have lost our government. The Constitution that our forefathers wrote--it had such meaning--is lost because our Congress is bought and paid for by the oil companies and the utilities. They're bought and they're paid for.

VANCE FISHER:

When that thing blew, I figured that all of the lawyers who were available to the power company were up at the capitol lobbying the state to prevent an evacuation order, trying to play the thing down, trying to keep it quiet, keep it cool, and I think they're still doing it. They're brainwashing the state officials or buying them off, whatever it takes.

MAYOR BOB REID:

Middletown has a very lucrative contract with Met. Ed. that was signed back in 1906 for one cent a kilowatt. There were some officials who approached me on the first day of the accident and said, "Maybe we shouldn't say anything about this whole situation. There's a problem down there at Met. Ed. Do you remember our

Pat Street with children, Jenny, age eight, and Mike, age ten.

contract?" And I said, "Well, the hell with you and your contract. I'm thinking of lives, my life, my kids, my family and other people here in town. To hell with the contract! What do you want, money or your life?" I just couldn't understand their thinking in saying, "Maybe we should stay out of it." I said, "Well, you stay out of it. I'm not!"

VICKIE DI SANTO:
It's not just us. There are other people who are going through the same problems. And always their problems were created by the needs of business—with business, anything goes and business will investigate later, and then it's too late.

It's very messed up. It is not people oriented at all. A lot of people wouldn't want us to slow down but we've got to somehow. We've got to slow down and start thinking about the people and what it's going to do to them, to the people, to the animals, whatever.

MAYOR BOB REID:
This is the thing that bothers me. The pro-nuclear people say, "We should just go ahead and have it." But, damnit, they should try to make it safe and not run rough-shod over the people. It's frightening to think that they say, "Look, let's just put a plant in and to hell with the people." Money, you know.

FRAN CAIN:
Met. Ed. has lied to us. I'm sure that if they continue to run this plant they're going to lie some more.

JANE LEE:
A representative from Met. Ed. was in our milk house after the accident when a T.V. camera crew was here from Washington, D.C. The reporter asked him how long he had been testing the milk here. He said, "Oh, we've been testing it ever since the plants opened." I happened to open the door just as that question was asked and I walked right up to him in front of the cameras and I said, "You are a damn liar! When are you people going to stop your damn lying? You have never set foot on this farm until the accident happened!" I was really furious at that point because these people have just lied left and right about everything.

MAYOR BOB REID
It's the same thing right now. I asked Met. Ed. about the release of krypton gas. I'm not for that. I don't believe that this gas should be released in the atmosphere, and I said to them, "If you do get the go ahead from the NRC to do it, I wish you would say when you are going to do it. At least give people who don't want to be in the area a chance to get their families and go somewhere. They might go to Reading and shop for a while, or Scranton." The company official responded, "Well, we don't know. We can't do it that way." I said, "Well, you just don't care about people then." They say they care about people but deep down inside, I really don't think they do.

FRAN CAIN:
Q: Has the accident changed your attitude about this country?
I would say yes it has. I don't think the government is taking care

of its own people. We have our noses in too many different things that are going on overseas like sending missiles or arms to foreign countries when we should be taken care of first.

Q: Has your attitude changed towards corporations and big business?

Why, yes. Let's face it. They're the money people. It's their say-so. They more or less rule the roost. It's the money that makes the moves, believe me.

Met. Ed. wants to raise fifty-five million more. And that will come back on us--the people who use the electricity are going to have to pay. Why should we have to pay it? It wasn't our error and we don't even get electricity right from this plant. But still we're harrassed by it. And we have the worries and the headaches of it.

PAT STREET:

I don't think I should have to pay for replacement power because I did absolutely nothing to cause Unit 2 to go out of service. I don't think I should have to pay to clean up. *I've paid enough.* I personally would like them to send me a check every month, at least quadruple what I have to mail them. I feel I suffer more by living here than they'll suffer by going bankrupt. They have in effect bankrupted us.

We find ourselves in the middle of these huge problems with no way of adequately solving them.

Q: Except you didn't create this problem.

That's right. I don't see why I have to pay for someone else's mistake, which is exactly what I'm doing. Me and thousands of other people.

GLADYS CONLEY:

When the accident happened, if it was a meltdown, we would not have been able to return to the land for forty or a hundred years. This has been our home. It is all that we have. What would have become of us if there was a meltdown?

The thing that really got me after this accident was reading some of the letters in the local paper by those who were pro-nuclear. They said, "It's perfectly safe. There was no danger. After all, nobody died. And, now, why condemn it? You think nothing of

54

going out on a highway with all the danger and everything." Well, there's no comparison because we take that danger on the highway upon ourselves. It's not forced on us like this really was. And we were here first.

PAT STREET:

Q: Do your children ever talk about it now?
Sure. Usually it's to say, "I don't know why they don't listen. Why do we have to keep telling them? I don't know why they don't listen to us." Jenny seems to think that they should listen the way kids are supposed to listen to parents. If something is wrong and you hit somebody, your mommy or daddy says, "Don't do that again," and you don't. She doesn't understand that you have to keep telling them and keep telling them. Michael just says, "I'm plain sick of it! Is that all you talk about?", even though it might be the first time in the day I've mentioned it.

My neighbors tell me, "Your kids are going across the road, sticking their tongues out and making faces every time a Met. Ed. vehicle goes by." I talked to my son about it. He said, "You know what's really aggravating is that they smile at me when I do that." So, my kids are hostile, they really are, more so than I am. As a matter of fact, I've been told I should take lessons on being mean from them.

We were at a technical meeting on the reopening of Unit 1. My husband was working so I had to take my children with me. My daughter was there drawing with her crayons on paper when one of the people from the NRC said that it was safe, there was nothing to worry about. Jenny looked at me and said, "Same old bullshit!" Actually I'm proud of them when you come right down to it, because they seem to have an insight to see straight through people. They're amazing. That's my Jenny, the little woman right there. She's the one that keeps me going. When my daughter and I spoke at hearings, she said, "My name is Jenny Street. I live less than a mile from Three Mile Island. I want Three Mile Island closed. I want it closed so I can live til I'm ninety."

I went up to speak and I began to cry while I was up there. I told them I felt like screaming for five minutes because we were limited to five minutes. But I started crying because I was telling them some of the things my kids had said, and how it had upset me. The

55

NRC official said to me, "Well, sit down and gather your thoughts. When you've got them together you can come back up and talk." He really aggravated me when I thought about it afterwards because he doesn't realize that it doesn't matter how much time goes by. I'm still going to be worried and I'm still going to be upset!

Afterwards, Michael, my son, said to me, "It doesn't matter what you say to them. They don't listen anyway. *They didn't even look at you.*"

Frustration

JANE LEE:

It takes a tremendous toll for me and I'm not getting any younger. I've gone the whole route, I have tried everything. People say to me, "Write to your congressman." I can show you a folder many inches thick of letters that I started writing in 1973 to my representatives in Washington, to the President, to anybody who would even listen. And I got absolutely nowhere!

I have sat in on all the hearings–all of them–and it was absolutely futile! The only reason they allowed so many hearings (and they had quite a few where people could testify) was so people would have a place to let off steam, satisfy themselves and walk away thinking, "Well, I got it off my chest, now I can forget about it."

I was supposed to be an intervenor on the hearings that are coming up. I sat through the pre-hearings for three days. After those hearings I realized how futile that was. I can't see the point. Let me tell you what they're putting the intervenors through. It's going to cost the organizations between thirty and fifty thousand dollars for attorneys and legal fees. If you are an individual, like myself, and you try to represent yourself, you'll have to pay all your own expenses. You have to make copies for all three people who sit on the licensing board, as well as the NRC council, the Met. Ed. council and all the intervenors. And you can imagine how many copies that are involved with all the contentions. Also, the NRC keeps writing back to these people asking them to rewrite their contentions. They have to submit all the names and the profession of the individual who's going to be a witness for them. Isn't that handy so that they can apply the pressure if they work for the state or for the Federal government?

The most interesting part of this whole thing is that what they're doing is litigating the entire proceedings before they have the hearings. They're requiring the intervenors to write and to rewrite their contentions. They are actually requiring us, in essence, to prove that we had an accident at TMI.

It's the most incredible thing and the public doesn't know what's going on—not a word of it has gone out to the paper. What they are doing is denying these people due process of law. They have eliminated me as an intervenor altogether. After the hearings I never went back. I never even bothered to submit my contentions. I never bothered to write to them or complain. I just figured the hell with it. It's so futile to try and deal with the NRC.

The NRC is no better than Met. Ed. The utilities and the NRC simply go to bed together. It's just what it amounts to and the intervenors are going to lose. They have let themselves be caught in the trap of trying to litigate something on the terms of the NRC. They should never have allowed themselves to be sucked into that trap.

Q: How could they have avoided it?

I think that what the attorneys should have done was to bring a countersuit in court and declare the NRC out of the picture as far as litigating anything because it is a conflict of interest. Instead of that, they allowed themselves, without realizing it, I'm sure, to be caught up in this thing. Now they're locked into it and there's no way out.

See, the NRC can litigate most anything because what they're trying to do is upgrade and change all of the proposals in the operations of the plant to make it acceptable. But what they can't make anybody accept is the report I have. Therefore, they had to find a way to eliminate me and that's just what they did. They eliminated any chance of my presenting this to the hearing.

Now there is an alternative to this. The one intervening group from State college can call me as a witness. But, still, they could tear me to shreds because I am not a qualified expert in their

Met. Ed. vice-president John Herbein at Middletown press conference, March 31, 1979.

opinion. So that disqualifies me, right there. But in my opinion, it is not up to me to prove anything. It's up to them to explain why these animals are dying and what is happening to the environment. And if it isn't their plant, then what is it? See, I don't say it's radiation, I don't say it's anything. All I say is, "Here the problem is. What are you going to do about it?"

FRAN CAIN:
I'm hoping they won't reopen but my honest opinion is that I don't think you can fight the U.S. government. I do not think you can fight the big money people and that's all what's behind this.

I got up in front of the NRC. When I go, it only makes me worse. It just stirs me up inside. I feel like I'm fighting a losing battle.

MAYOR BOB REID:
The government says we need it. Ninety-nine per cent of the people in this area could holler and scream that we don't need it, we shouldn't have it, but Uncle Sam is going to do it, anyway. They have played this game for so long, for almost 200 years, telling the people, "Yes, you're right, you have a say, but in this situation we think we know what's best for you."

It's not democratic. But they have done it for so long that we have accepted it. I really think that it's a losing battle. I think the government has the people where they want them, with the energy crunch and with what's going on in the Middle East. I think it's in their favor.

PAT STREET:
It's frustrating! I feel like I'm banging my head against the wall. I really do. And I try to be nice. I have broken my back trying to be nice. But it does not matter if you stand up there and talk calmly and make sure you don't insult anybody. Or if you stand up there and scream and call them every name in the book, they still won't listen to you. If we could get them to put at least ten per cent of their emotions into any decision they made, it would be in our favor. But they don't have emotions. I told one of the attorneys for Met. Ed. that I wish Bob Arnold (V.P. of Met. Ed.) would stop smiling whenever he answers a question. I said, "I noticed that he

was smiling when somebody was asking him about the health of their children. It really bothers me that he can stand up there and smile, no matter what you ask him. That isn't P.R. to me!" It's just *outrageous* because I tell you I very rarely smile these days and I don't think he should be able to. They're telling us that they apologize but they sure do smile a lot. I'm thankful that at least I have two kids and have reason to smile.

Loss of Trust

VICKIE DISANTO:

Before the accident, I was very pro-nuclear. My husband and I always believed in technology, in advancing ourselves through it. We were very proud to live down the road from a nuclear plant—we thought this was very advanced. People would say to me, "How can you live so near TMI?" I'd say this was modern technology and we've got to have these things. The government assured us everything was fine and I trusted our government. I believed that the nuclear industry and our public utilities were serving the public. We were sure that nuclear power was safe. And we tried to convince ourselves of that up until the time that we left our homes. But living through the accident with the thought that we might never come back here really hit home and made us realize that it's not safe. There's a lot that we don't know and a lot of risks that haven't been considered.

Q: Has the accident made you more aware of a lot of things that are wrong with this country?

Oh, yeah, because any time anything happens we worry for other people because we know what it's like. So if I hear about some tragedy for someone else--like Love Canal--it makes us remember what we have gone through. Of course, we tend to side much more with the victims. Whether it's merited or not, we blame just about everything on government and big business.

Everything has changed for me. I don't believe there's anything we can trust the government with.

Q: And has your faith also been shaken in big business?

I have no faith whatsoever in big business. Just about every business I've ever seen was corrupt.

Q: What changes would you like to see in this country?

I think we have to be a little bit more concerned for the individual and humanity as a whole. There's too much concern for money. I don't think that's fair--it's the individuals who make up the country and everyone seems to forget that.

FRAN CAIN:

Q: Did you have any feelings about the plant, the dangers involved in it, before the accident?

Not really. We were told that the plant was one hundred per cent safe and we would never have anything to worry about. They did mention about low radiation leaks with us living so close. But they said it would not hurt us or affect us in any way. Generally, I had a lot of trust in the government. I've always thought the U.S. government was for the people.

Of course, since the Nixon Administration, you really can't trust the government. Whether they're Republicans or Democrats, I feel that if they are straightforward men when they first go in, they soon learn the greed for money and power.

Q: Was it Watergate that first made you feel differently about the government?

I believe to some extent it did, yes, but not nearly as much as since this TMI accident. The reason I feel this way is because the TMI accident has affected my life to a great extent.

MAYOR BOB REID:

Q: Were you concerned about the plant before the accident?

No. Being a layman, I had to put all of my trust in those people. They assured us that there would never be an accident. The only thing I knew about the atom was the darn bomb. So these people said everything was going to be all right and I believed and I trusted them. I don't believe them any more!

Q: How has the accident changed your life?

It's going to make me more concerned with anything they tell me is all right. In fact, I know quite a few people who feel this way. They're going to question a lot of things, more things than they did before, question everything--is it really the truth?

Fran Cain stands in front of her dog-breeding and grooming business holding her eyeless puppy, Kelly. She operates her shop at her home which is 1,000 feet from Three Mile Island.

JANE LEE:

Q: What motivated you to begin your one-person investigation of farm animal problems in the area?

Well, before I did the report I got involved with the township when they tried to implement a landfill. What I discovered was the township and the state had lied to us continuously about the geological surveys that they were doing in preparation for this landfill. When I found out that the geologists at the Department of Environmental Resources had not even read the engineer's feasibility study, and that the survey team, sent from State College to do the geological examining of the soil for proper leaching, was in error, I knew that we had been lied to.

When we defeated the landfill proposal, I realized that there was a strong possibility that there was a lot wrong at TMI. That's when I began my investigation. I began by writing to my representative in Washington. He sent me all of the hearings and a book the size of a telephone book on nuclear power plants. I sat down and studied them. From that I went into radiation and genetics. I read several articles and books critical of nuclear power written by some very reputable doctors. That really opened my eyes and made me realize that we were in serious danger.

The On-Going Accident

AN INTERVIEW WITH DR. JUDITH JOHNSRUD:
We knew from the day, virtually the hour, the accident was first announced, that it was a very serious event and that the accident would continue indefinitely into the future. Only now are the people of central Pennsylvania beginning to understand how serious it is and how much worse it's likely to get in the future. And so, as the accident continues to unfold, I think people elsewhere in the United States, because the news media lost interest, have not understood at all the nature of the on-going problems at Three Mile Island. And not until these problems become critical again will the press pay any attention. Day by day by day, the NRC is allowing things to get worse at Three Mile Island. Met. Ed. continues to exhibit its total irresponsibility toward public health and safety, and the people of Pennsylvania are left to suffer. Indeed, while many of the officials may say they think there's no problem, underneath they know that the nuclear industry is causing irreparable damage to people's lives.

Q: What are some of the specific problems that do exist with the plant now?

In a number of ways the accident is still in progress. First, of course, is the problem of releasing krypton gas into the atmosphere. Krypton has a half-life of eleven years, so any that is released will float from Pennsylvania up to Massachusetts and beyond. And somewhere, sometime, someone will experience a small dose of radiation from the krypton that is released. The people most at risk are the people here. Now, the industry tells us that it's a small dose--insignificant, but the point is that it's an additive dose above doses that we don't know about because the data isn't available from those first days of the accident. So it's just a little bit more that adds a little bit more to the absolutely intolerable risks to which the people here have already been exposed to.

Secondly, the release of krypton on the grounds that Met. Ed. has put forth will set the precedent, which in the future, Met. Ed. will refer to in order to be allowed to release the other radioactive contents of that damaged reactor into the environment.

The NRC has in its regulations a criterion that is used to determine how much release may take place from a plant. Rather than having hard firm numbers that may not be exceeded, they have a criterion they call ALARA which means as low as reasonably achievable. Now let's think about what "as low as reasonably achievable" means. If it's your life that's at stake, you may have a different notion about what's reasonably achievable than a utility. And so the NRC has put a price on your head and on the heads of us all. That price is one thousand dollars per person rem. So, if it costs the utility more than one thousand dollars per person rem to reduce the exposure, then it will be allowed to proceed, no matter what the damage may be to the public. And so, Met. Ed. has said that it's too expensive to do anything other than vent the krypton. They can do other things. They have a whole containment sitting over at Three Mile Island--Unit 1, the undamaged reactor. It's off-line. They can store it there.

The next way in which the accident is still in progress pertains to the large quantities of radioactive water that remain on site. There are two bodies of water. The first is the intermediate level water in the auxiliary building which is being treated with what they call the Epicore Two decontamination system. We've only recently learned that the Epicore System is not performing in removing the bulk of the radioactivity as was promised. Met. Ed. is now admitting that they went ahead and constructed it without the capability of easily solidifying the highly radioactive resins that are left which collect the contaminents.

The second body of water, which concerns us perhaps the most, is the highly contaminated seven hundred thousand gallons of water sitting in the basement on the containment building at TMI.

Now, that concrete was not constructed as a swimming pool. Yet, it's been filled with hot water for over a year. There are concerns about seals, pipes and potential leakage. If it leaks, it's the Susquehanna River and the Chesapeake Bay that will be contaminated.

Now, let's say that Met. Ed. successfully tanks up the krypton rather than releasing it. Let's say that they successfully decontaminate the water rather than letting it be released. Then we have to go into the containment building and begin to scrub it down. The walls are contaminated. They are permeated in those lower layers with the cesium and strontium contaminated water. So we can anticipate that some very unpleasant contaminents will be found in the dust that's created as they start to clean up. And if they are allowed to use the "as low as reasonably achievable" standard for filtering that dust, we can expect some long-lived radioactive fallout from the dust particles will be released over the rich agricultural lands of eastern Pennsylvania. That's the third way in which the accident is still in progress and we believe it is going to get worse.

One of these days, the workers of Met. Ed. will have to take the head off the reactor and go in there and try to clean it up. That is where the worst problem lies ahead of us. The president of General Public Utilities (parent company which own Met. Ed.) has said in a letter to the NRC that the utility now fears recriticality of the core. Now there are several ways in which that can happen. One way would be through the loss of sufficient boron in the water, which is presently keeping it under control. Another way this could happen would be by an uncovering of the core, which of course still has decay heat coming out a year later. They have to keep that core covered. If they have an uncontrollable leakage, as they have suggested may be the case from the reactor, they may lose the covering and that in turn could lead again to a heat-up of that core.

We do not know the condition of the fuel in the core. There's general agreement that there has been a crumbling of a lot of the fuel. One of the NRC technical people described it as having crumbled to a consistency somewhere between popcorn and granola. And further, we don't know the condition of the control rods in that reactor core. And so, when the head comes off, the last barrier between that still potentially lethal beast and the public, will stand

68

exposed to the experimentation, which is the only thing to call it, that will take place in trying to clean up the interior of that reactor.

The way ahead looks very bleak in terms of public safety. We're very deeply concerned about it here in Pennsylvania. We want the rest of the world to know what a nuclear reactor accident is really like.

Q: If a pipe breaks in the containment building and the core gets uncovered again, is there still danger of a possible meltdown?

From what we've been able to gather from NRC technical people and from our own technical experts' work, the answer to that is yes. It would not be a meltdown in the same way that we were so concerned with at the beginning of the accident. Nonetheless, the slow heat-up in that core could result in further melting. It would take a longer time. I think the question is: what would they do about it? With totally inoperable safety equipment, they no longer have those famous "redundant" safeguard systems that we've been promised would take care of us under any circumstances. The truth of the matter is that neither the designers of nuclear power plants nor those who construct them nor those who regulate them have the intellectual capability to think of everything that could go wrong.

They certainly never thought of this or if they did, they ignored it. Any way you look at it, it's the public and future generations who will pay the price for their arrogance. I can't call it anything other than *criminal* arrogance.

In the event of a breach of the containment structure, the area that would experience deaths from the radiation release extends as much as ninety miles downwind. In fact, at the time of the TMI accident, those living in Massachusetts would have received, according to the AEC's own estimates fifteen years ago, doses in excess of ten rad. And so the areas that would be damaged with resultant premature deaths associated with an accident are enormous. It was the AEC that said fifteen years ago that *an area the size of Pennsylvania* would be contaminated. It wasn't the screenwriters for "The China Syndrome."

I'm very troubled by the events surrounding the reactor. I'm at least as troubled by the attitude of the NRC. They have learned lessons from Three Mile Island all right. Their lessons are not to improve their licensing and siting of reactors, nor to improve the understanding of the hazards of nuclear energy for the public. Quite the contrary, the lesson learned by the NRC in the accident at Three Mile Island is to keep the public ignorant. To restrict those who would try to protect the public interest by intervening in the NRC's licensing proceedings. To use steamroller tactics to continue what I feel to be morally reprehensible actions, mainly the continuation of the nuclear industry.

I can see no justification whatsoever for a single kilowatt of electricity from nuclear generated facilities, not with the price that has been paid by the people of Harrisburg and the people of central Pennsylvania. The damage to them physically is far greater than the NRC is willing to admit. I say this because the NRC does not know how large the original doses actually were. During the first three days of the accident at Three Mile Island, there were no ground-level doseimeters between 2 ½ and 9 miles of that plant. And so how large the doses were to that population we will never know. However, the figures that the NRC has admitted, in terms of releases from the back calculations and from the monitoring that has been done, simply doesn't add up to a mere eighty-five millirem less than a year's background exposure. The figures indicate that the doses must have been substantially larger to the people who live close to the reactor.

The people here are living with the uncertainty of how large a dose they received. They will never know. They will never be compensated. They are watching their children become ill with hyperthyroidism. They find infant mortality rates that are higher than expected. Yet, they are told that it cannot possibly be from the reactor accident. There is no way, that they, the victims, can prove otherwise. That's part of the tragedy of Three Mile Island.

In the event of a loss of the containment structure, those people who live within a twenty mile radius, even if they run for their lives, don't have a prayer of survival. The reason for that is perhaps the most outrageous of the NRC's changes since the accident. Now, in accordance with a NRC document, numbered NUREG 0610, the utilities will follow new guidelines in announcing the next nuclear accident to the public. The NRC is allowing the utility to wait to declare a general emergency, with hazard to people offsite, until more than one thousand curies of Iodine 131 or its equivalent have been released, more than a million curies of xenon 133 or its equivalent have been released and until core melting is *in progress with loss of containment imminent.* By that stage of a reactor accident, it will be too late to run. There will be no evacuation. That's what they are really saying to us.

Q: Are these new figures for an accidental release of radioactivity quite high?

Yes, Iodine 131 is particularly hazardous in terms of inhalation and effect in the thyroid as well as ingestion through milk. The NRC estimates that only fourteen curies of Iodine 131 were released. Yet they are now saying they would allow a *thousand* curies of Iodine 131 to be released before thay announced a general emergency. I think this is a total abdication of the NRC's responsibilities to protect the public health and safety.

Q.: How do you feel about the future?

The Three Mile Island Reactor Number Two sits there in the middle of the Susquehanna River like a wounded, angered beast. And we never quite know, in its death throes, how it's going to lash out at us next.

Krypton

On June 28, 1980, Met. Ed. began venting 57,000 curies of krypton-85 radioactive gas into the atmosphere from the containment building of the damaged Unit Two Reactor at Three Mile Island. Within four minutes, alarms sounded on all the radiation monitors surrounding the plant. As always, Met. Ed. was quick to assure the public that there was no danger and insisted the problem was due to faulty alarms.

The venting was continued the next day and was completed ahead of schedule, two weeks later on July 11. The krypton venting took place as a part of the clean-up process at Three Mile Island.

The following interviews were conducted several weeks before the venting took place.

FRAN CAIN:
When Met. Ed. Vice-President Arnold sent Mr. Schneider (a representative of Met. Ed.) in to interview me, I was extremely upset. I told him I wanted to move. I said, "You are going to release the krypton, aren't you?" He said, "Yes, when the weather and the wind are right." I said, "How will the wind and weather ever be right? Because it has to go down." The man turned around and said to me, "We will release it on a windy day but you don't have to worry because when it goes up the majority of it will land down

72

right here on the island, anyway." *So how do you believe a company when they tell you this?*

I'm very much against the venting of the krypton. I'm trying to get the Met. Ed. representative to help me with transportation for my dogs when they do vent. But I don't think they're going to do a thing for me. If it was just myself and my child and two or three dogs, I would take off and go. But I do not have the means to take twenty-five dogs with me. I really need a trailer to get them out. The last time I spoke to the representatives from Met. Ed. about the animals, he said that Met. Ed. shareholders had taken such a loss that they feel that they just don't have the money to do stuff like this. They're the ones who have presented the people around this area with this situation so I feel that they're the ones who should help us.

BILL WHITTOCK:
Q: What would you do if they vented the krypton?
Well, if they'd told me that it was coming my way, I'd get the hell out. And I guess a lot of other people would, too. If they were venting krypton and the plume was coming over my house, I'd get out. I'd be foolish not to, I don't want to get into a plume of radioactive gas.

PAT STREET:
Q: What will you do if they vent the krypton?
Probably I'll go to my mother-in-law's with my kids in Maryland.
Q: If you go, will your husband stay?
He'll have to! Do you think they're going to give him a two-month leave of absence with pay? No way. This will be one more strain on the family.

VICKIE DISANTO:
Q: What will you do if they vent the krypton?
I'll probably take my children and go stay with friends in New Hampshire.
Q: Would your husband go, too?
You see, that's the problem with leaving. It would not be an ordered evacuation. So my husband would not have work if he went with us. We'd still have to keep up all the payments on the

house, plus our living expenses wherever we'd go. Money is generally pretty tight for us.

MAYOR BOB REID:
I feel that the people should have a say since it's their lives. They should hold their own fate in their own hands. I think through some kind of a vote or survey, the people should have a say. I don't think it should be up to a small group of people to say, "Go ahead and vent it." I think the people themselves should have a say. Remember it's the people's health and lives that the utility is dealing with. But I know that the utility is not going to do that.

On March 19, 1980, the NRC held a special public hearing in Middletown to inform local residents about their proposed recommendation to the Met. Ed. Company concerning the venting of radioactive krypton gas into the atmosphere. What follows is an interview with Cathie Musser who attended the meeting.

CATHIE MUSSER:
We tried to get to the NRC hearing early. So we decided we'd arrive about ten after seven. The meeting was scheduled to start at seven-thirty. At ten after seven they said they weren't going to let any more people in and closed the doors.

It was quite obvious there was standing room inside the building: we could see inside. The policemen were trying to limit the number of people to the number of seats. But people wanted to stand and were getting really mad. So, someone opened a door and a lot of us got in and stood around the back of the room. Then they locked the doors. It was really bizarre because there was about two hundred people outside the doors banging on them. I found out later that some of the louder bangs were coming from a New York Times reporter who wasn't allowed to come in to attend the meeting. The NRC had barely got to open their mouths before the people started screaming and hollering. They didn't want the krypton to be released and they were sick and tired of decisions being made without their input. Even though the NRC was saying this meeting was to hear the public comment and that no decision had been made, the general concensus was that the decision had already been made. There's a distrust here of the NRC which started a year ago during the accident.

The NRC tried to put across their views of the different ways that they have considered to get rid of the gas. They were barely listened to. It was a very, very angry crowd, something I've never seen around here before. There was a lot of screaming and yelling. It did settle down for a while because they held a very orderly question and answer period where you had to come up to the microphone and give your name and then ask your question or make your statement. So that stayed relatively calm except when someone made a statement that got everybody riled up.

Q: What kind of statements did people make?

Basically they said they did not want the utility to vent. People were saying, "We want to leave if you vent, no matter what you say

Union Street in Middletown in July, 1980.

about it being safe." It was the angriest I've seen people since the accident.

Q.: Why were people so angry?

I think that up until this point they thought that the amount of radiation that they received would be it and that the NRC and Met. Ed. wouldn't dare give us any more. They were saying, "You mean to tell me that you are really going to do this again no matter what we say or think?"

People are just basically tired of being afraid, tired of having to listen to the radio every hour on the hour to make sure something else isn't going wrong at the plant. It's been a long, hard year for people to live with an undercurrent of anxiety. When they start talking about venting more radioactive gas, then the anxiety comes to a head in a real hurry. It's been building up. I've seen some angry meetings, but nothing like last night. I was really shocked.

One local resident asked one of the NRC gentlemen where would his wife and children be during the venting. And he replied, "Well, I would be happy to bring them here." Of course that brought all kinds of boos and hisses and hollers. But then this gentleman from the audience said, "Would you please jot down this address?" He gave him an address. The NRC man asked him what it was. The man answered, "It's my home. I'm not going to be there so you and your wife and your children may stay at my home." Another father stood up and said that he was going to send his wife and their two children away and would the NRC man pay his expenses to go visit them, up in New York, for two months. There were some people who stood up with technical information. Some mentioned the A-bomb tests in Utah during the fifties, "They said it was safe then and now they are finally admitting that it wasn't. We feel this is the same thing you're doing to us. You're saying it's safe now and twenty years from now you'll be saying that it wasn't."

It was a very mixed crowd, not at all the radical type. It was mostly middle class people. This is what was so shocking, to see these people acting like that.

Q: Are people feeling really frustrated?

Yeah, frustration is probably the biggest part of it. I feel that it certainly doesn't make sense to keep quiet, although some people are taking that option for the simple reason that they can't deal with the frustration and the anger any more. They decided to try avoiding the issue to see how that works.

I found it doesn't work. I find that I feel a whole lot better when I'm out actually doing something about the problem.

Q.: What do you think will happen if the NRC does decide to release the krypton?

I'm sure a lot of people will leave. But we're not talking about a week or eleven days, we're talking about almost two months. And you cannot pick up and leave for two months without going through some real hardships. People aren't going to have jobs when they come back, which is what a lot of people are concerned about.

What I'm concerned about is our daughter. My daughter is only two years old. It's one thing to stick around and maybe get your extra millirem or so when you're an adult, but I feel very strongly that a child needs none or certainly as little as absolutely possible. Since nobody really knows how much she got last year, I don't feel comfortable having her get any more. So that's what most people are going through. They want their children to go away which means the mother has to go with them. So you're splitting up families. The father stays home and tries to save his job. The wife and kids go off and make themselves a burden to a relative or a friend for two months. That's a big decision to make. I've considered maybe going a hundred miles away or so and camping so that my husband, Jim, could come up and visit us. He wouldn't be able to visit us maybe except on weekends. That's a little bit like having him go to war and come home on leave.

We lived five miles from TMI when the accident happened. Now we live about twelve. I don't feel real good about it. I'm really angry, as a mother, that my daughter has this hanging over her head at such a young age. I wish it never happened, and I hope that it never happens to anybody else. I hope that they learn from this mistake and start backing down on nuclear power and phase it out. Let's get rid of it.

Even if, twenty years down the line, they prove that there was no health problem, the psychological pain that we've lived with is a big problem in itself, the terror we went through during the accident, and I mean *terror*. It was awful. Just fleeing, not knowing whether you were at that moment being bombarded by radiation. Then sitting a hundred miles from home and watching television and hearing them talk about a hydrogen bubble that might explode and uncover the core. They were talking about my home. I might never be able to go back there! That was unbelievably terrifying. And what has followed has been sort of a low-level terror of an underlying nature. It really eats at you.

Never Again

FRAN CAIN:

I realize that we need energy. I feel that there must be a safer way to go than what we have right now. I can never trust nuclear energy again because it happened once and there's no saying that it's not going to happen again. I was here before this was built. It's just like they've taken ten or twenty years away from my life.

Q: What would you say to people who live near nuclear plants in other places around the country?

Well, it might not happen again like it did here but to me there is always the chance that it would. If they could fight to have them closed down, they're better off than to have something happen like what we've experienced here. If a meltdown comes you've got to go six hundred or seven hundred miles away. But we're sitting right here and they say *evacuation?* I do not honestly believe that they could evacuate all these people in time. There are just too many people living around this area.

Q: If a meltdown happened or a major release, you're so close it would almost be too late to evacuate.

That's right. We'd have very little chance. If it had been a meltdown, you couldn't come back to the land to begin with. You would have lost your life's savings, your life's work.

Now, there are a lot of people who don't feel like this. I know my next door neighbor tells me, "You could die in an automobile accident." That's right, and I know when "the man up above" is calling me, I'm going to go, but "the man up above" also wants you to help yourself.

79

Q: Has the accident changed your life by making you a more public person?

Yes, it definitely has. I feel I have to stand up for what I believe in. And this power plant--this monster that's sitting across the street–I'm very much opposed to it. I don't feel they're safe at all. I think that the "man up above" is very angry and has been trying to give these people a warning, telling them that this can happen again. And the next time I feel that people may not be as lucky as we've been thus far.

PAT STREET:

Before the accident, my main concern was getting to work, getting home in time for the school bus, getting supper on the table and doing the dishes. I was worried for the people that worked there but I wasn't worried for me. And I wasn't worried about my children or my husband or anybody else who lived near the plant. I thought it was basically safe.

I've done a complete turnaround. Now, I'm worried about *everybody*, not just the ones who live here, by the way. I'm worried about people who live near other nuclear sites, too.

I want my family to live. Since they're my number one priority, the island *has* to be closed.

I saw a TV special on uranium mill tailings where there were Indian children playing on a pile of them. That made me sick because I didn't realize things like that went on. I realized that they are probably in a lot more danger than we are. And to hear the officials say that it was low-priority in terms of safety because it was on an Indian reservation–I was just about ill!

Q: Has the accident changed your attitude toward government and big business?

Definitely. It's over a year later and no one has listened to us yet, even though we're in the majority. I thought democracy worked by majority rule. I guess it's only if the government agrees with you.

Q: How will your life be different after the accident?

I will be active in the anti-nuclear movement until they're all shut down, because no one is going to have to go through this again, if I can help it. I'm only one person, but it has to be done.

I'm not a public person. I'm basically fairly private. This has changed me. I've been on a television show and I spoke at a rally. I would never have dreamed that I would do anything like that before. And I'm not totally self-assured when I do it. I get upset and I'm nervous, but I do it anyway because I feel that people have to know what's going on. The nice thing about the anti-nuclear movement in this area is that the people come from all strata of society, from housewives to businessmen to blue collar workers. The age group is just as varied. For instance, my daughter testified at the NRC hearing at age seven and there were people of seventy who also spoke.

Q: What effect do you think this accident will have on the anti-nuclear movement?

I hope that it will make people work harder. It's a lesson we should all learn from. Shut them down before they shut down humanity. We've got to take our message to all layers of society, take it into churches, women's groups, business groups. . .in plain words, it means an awful lot of work.

DR. MICHAEL GLUCK:

Before the accident happened, I really wasn't involved with politics. I was never really a "cause" person or anything like that. I was fairly apathetic. I was going through a lot of med. school, putting in a lot of hours and doing a lot of work. It made me pretty apathetic to a lot of things. But now, I have two children and I feel responsibility to make the world a better place for them.

Q: How has the accident changed your life?

It's made me more aware and more responsible in trying to shape my future and my kids' future. There are certain things that have gotten out of control. I felt I was placed in a special situation as a physician. People tend to listen to physicians, so I felt it was an opportunity for me to try and do something to stop what's going on. This is really the most important issue facing us today.

Q: What has been the reaction from other doctors to your activism?

Most doctors I work with have been supportive with what I've been doing, but they haven't been active themselves.

Dr. Michael Gluck, General Practioner.

As far as my role, I feel that I can try to educate the people around here about the dangers of nuclear power, what it means to their health and their children's health. It is a fact that a normally operating plant may increase the risk of cancer and leukemia as well as increase the problems in pregnancy and the number of stillbirths, etc. As a physician in this area that's about all I feel capable of doing I try to speak at any meeting I'm invited to, such as the P.T.A. and other community meetings and rallies around here, where you get to talk to the grass roots people.

The people involved in the movement down here are really from all social classes, all walks of life. There are people who were totally trusting in the government, people who supported Richard Nixon and Spiro Agnew, who are now leaders in the anti-nuclear movement. These are the people who are the most active in the cause, people who served in the army, did their duty and really supported the country without question. These are the people most deceived and most hurt by what happened. A lot of people totally trusted corporations and the government and feel totally deceived by them all now.

Q: As a doctor, what do you tell people about nuclear power?

We know that there's increasing evidence that living near nuclear power plants when they operate normally can increase the incidence of leukemia, cancer and birth defects in children. In the event of a major nuclear accident, if you live close to a plant, the chances of being notified in time are nil. The chance of getting out in time are nil. Evacuation is impossible. The plants that are built near cities, such as Indian Point, north of New York City, have no adequate evacuation plan. The Department of Civil Defense says that in the event of a major catastrophe they can move eight hundred people out of New York City's population of ten million.

I feel that the risk of obtaining thirteen per cent of our electricity from these plants is just enormous. You're dealing with millions and millions of lives, millions and millions of acres of land. It really is posing a tremendous risk for people in order for others to make profits. My morality says that it's not right. That's essentially why I'm doing what I'm doing.

ELIAS CONLEY:

If there's no danger in it, I say the best thing they could have done with it would have been to put it in Philadelphia. Why would a man be crazy enough to build a one hundred mile line to run the electricity back there? If they're safe, why then they ought to make the ones that work there and the ones that run it, live right around it.

MAYOR BOB REID:

Well, like they say, it's progress. But I don't like that type of progress where you're possibly going to cause the death of a lot of people. But the type of accident you have here is a different type of accident than with most energy related things. I'd rather see more coal produced. People say, "You have accidents with coal."

Yes, you have accidents in mines. Sure. But, generally, accidents affect those people who work there. If you're working around a big boiler and it blows up, it only injures those people in the immediate area. If you have a mine and it caves in, only those people are affected and they watch out for it all the time. But when you have the potential to kill a lot of innocent people, it's a different story.

Q: Do you think in a meltdown people would have time to get out?

In a meltdown you can forget about evacuation. I don't care what kind of evacuation plan you have. You can forget about it.

I went to Hiroshima last summer for the anniversary of the atomic bombing on August sixth. I saw the effects of acute radiation. It's frightening as hell to see how one bomb wiped that whole city out.

I wasn't worried about the bomb. The explosion didn't concern me. But the radiation concerned me because I could associate the radiation there with the radiation here from the plant. I went through the museum over there and saw the effects of radiation on the people, skin just falling off. The curator of the museum—his fingers, his hands, his face were deformed. Radiation is some rough stuff. I know if I would have been in Japan at the time of the dropping of the bomb and saw the effects, I'd be screaming to the rooftops against nuclear energy.

I'm neither pro-nuclear nor anti-nuclear but I have more tendency to lean towards anti-nuclear. I really feel that the fewer plants we have, the better off we are. I feel that the states that have an alternative source of power should use that alternative source. States like California, where they have a high rate of earthquakes. . .Boy, to build a plant out there you'd have to be sick. And for the people to *allow* it to take place, they have to be sick! If I lived in California, I would be raising hell every time they even talked about building a plant. This business of saying, "We can build it strong enough to withstand an earthquake." They're crazy!

I addressed the New York City Council. I told them the people responsible for building the plant there at Indian Point. . .man, they ought to take them out and shoot them! Thirty-four miles from seven and a half million people—to construct a plant with the winds always blowing towards Manhattan! Isn't it stupid! They asked me about an evacuation plan. "Man," I said, "forget it!" You got two tunnels and three bridges to move seven and a half million people."

I went down to Louisiana and gave a talk to a group of builders of nuclear plants. They came to the meeting just to hear me. (laughs) Money people, too, man. They were money people and there I was up there. And I told them, "You people are crazy down here to allow a nuclear plant to be constructed. You got enough oil, you got enough natural gas. If you strike a match in the wrong place you'll blow up the whole state." Boy, after the meeting did they come up and give me hell. Oh, man, did they give me the devil! And I told them how I felt. I said, "That's how I feel, mister. You with all your money. You with all your lies. I'm not going to buy it. Now you can tell these people down here what you want, but I saw it, I saw what happened. Now you can lie to these people but you can't lie to me."

JANE LEE:
Unless there is a very strong change of attitude in the scientists who are sitting down there in Washington, the nuclear power problem is going to be solved in the streets because this is not the last of the accidents. I keep saying in the back of my mind that if there's ever to be a demonstration it has to be women and children. There isn't anybody in this country who is going to turn a gun or a water hose or anything else on children. I think that the women in this country have to take their children and walk and stand in front of that plant and say, "No way are you ever going to turn this plant on." Now what are they going to do with children standing around? Is there any man who's going to have nerve enough to shoot a child? I think not. With women standing with babies in their arms? I think not.

VICKIE DISANTO:
Q: If you could give a message to folks who live around nuclear power plants, what would you say?
Get out while you can. It seems that no place is safe enough. But I say get out of there. Because we didn't get out and now we can't. Even if we do leave, it's still there. It's already affected us.

JANE LEE:

Shut them down before it's too late. Amen. Shut them down before it's too late, any way you can, but shut them down. Get those plants shut down because if you don't, you're going to be paying the price just like we're paying here. You're never going to have any peace of mind as long as those plants are running. Shut them down. Shut them down. Any way you can, but shut them down.

VICKIE DISANTO:

It's important to us how we feel. I think it's going to be important to a lot of other people how we feel. Or should be. I'm hoping people will pay attention to how people feel here. We know what we're going through. And everyone here knows what we're going through, because we're living it for everyone. Nothing's going to make it worthwhile for us to have gone through it, but if we can keep anyone else from going through this same thing, it's going to mean something. It will ease the pain a little bit.

DATE DUE